PLANTS FROM TEST TUBES
Revised Edition

PLANTS FROM TEST TUBES

An Introduction to Micropropagation

Revised Edition

LYDIANE KYTE

Foreword by Hudson T. Hartmann
Preface to Revised Edition by Ralph Evans
Illustrations by Sandy Godsey

TIMBER PRESS
Portland, Oregon

ISBN 0-88192-040-1
Printed in Hong Kong

Timber Press
9999 SW Wilshire
Portland, Oregon 97225

Library of Congress Cataloging-in-Publication Data

Kyte, Lydiane.
Plants from test tubes : an introduction to micropropagation
Lydiane Kyte ; foreword by Hudson T. Hartmann ; illustrations by
Sandy Godsey.
p. cm.
Bibliography: p.
Includes index.
ISBN 0-88192-040-1 : $24.95
1. Plant propagation--In vitro--Laboratory manuals. 2. Plant
tissue culture--Laboratory manuals. I. Title.
SB123.6.K98 1987
631.5'3--dc19 87-24607
CIP

Contents

ILLUSTRATIONS

TABLES

Foreword

The field of tissue culture micropropagation is almost explosive in its rate of advancement. Scientific plant science journals are filled with articles describing aseptic *in vitro* propagation of more and more kinds of plants. Earlier these were all herbaceous types of plants but woody plant species are becoming more commonly propagated in this manner.

In the nursery business, the plant propagators, geared to using the traditional propagation methods—seeds, cuttings, grafting, budding, division, etc.—found themselves in a quandary. Were they being left behind with their conventional methods in this rush to high technology scientific propagation?

However, these new aseptic propagation methods have been around sufficiently long to show that they are here to stay as a new tool for those propagators with the inclination to use them, and the funding for the equipment and for the necessary supplies.

Along with this, however, the plant propagator needs to know what to do, either on his own, or with the aid of trained technical help. Fortunately, much has been written about tissue culture micropropagation—scientific journal articles, reports of tissue culture symposia, scientific books, and a few books on the methodology involved, written for those with little scientific background. The present book by Ann Kyte gives a detail that insures success if diligently followed. It meticulously describes the facilities and equipment needed and tells what to do, step by step, in propagating a range of plants by tissue culture micropropagation.

Whether to adopt these new techniques for commercial propagation or to use the traditional methods depends upon the individual situation. Even though certain species can be propagated by tissue culture methods, the conventional systems might be the most economic to use. In other situations, and with certain species, production of new plants in massive amounts is possible only by aseptic micropropagation methods. These situations are well described in Ann Kyte's book.

Hudson T. Hartmann
University of California at Davis

Acknowledgments

This book was born out of a personal need for such a book when I started into tissue culture. It is, at best, a starting point and, hopefully a partial answer for those who ask, "What is tissue culture? As a grower, is micropropagation something I can and want to do? And how do I do it?"

I would like to acknowledge with gratitude:

Dr. Hudson Hartmann (U.C. Davis) for so generously reading this work, providing useful suggestions, and writing the Foreword.

Dr. Dale Kester (U.C. Davis) and Gayle Suttle (Microplant Nurseries) who so kindly read the manuscript and assisted with helpful comments.

Bruce Briggs (Briggs Nursery) and Sarah Upham (Native Plants) and countless others for their encouragement and willingness to share hard-earned information.

Sandy Godsey for her inspired and carefully executed drawings.

Richard Abel, editor, for his patience, reenforcement, and untiring effort to help make this book a credible reality.

Bob, my husband, for his support, assistance, and encouragement.

L. K.

Preface to Revised Edition

During the last decade the use of tissue culture in propagating many important species has increased dramatically. Today the advantages of tissue culture techniques for plant propagation are recognized worldwide, with millions of young plants being raised "in vitro" every year.

While *Plants from Test Tubes* serves as a primer for those nursery people wanting to understand the new technology and its applications for the horticultural industry, it is also a valuable reference for anyone involved in tissue culture production. It is easy to read and contains lots of practical information just right for the student and with sufficient detail for hobbyists to get started on their own. I have recommended it to many students, nurserymen and hobbyists wanting basic information on the process of plant tissue culture, at an affordable price.

This revised edition, updated and expanded as it is, should please an even wider readership!

Ralph Evans
Past President
Twyford Plant Laboratories, Inc.

SECTION I

Fundamentals

Chapter I: Introduction

Tissue culture is an important new method of plant propagation now readily available to growers but not widely used. It is time for this amazing process to step out of research and science fiction and go to work for the progressive grower. When a piece of a plant is placed in the tiniest of greenhouses, a test tube, it will appear to perform miracles. In the absence of microorganisms and in the presence of a balanced diet of chemicals, that bit of a plant will produce tiny replicas of its single parent. It will produce the replicas in incredible numbers, so numerous that they must be divided repeatedly and frequently to survive.

When plants are multiplied vegetatively, whether by tissue culture or by more conventional means, as distinguished from growing from seeds, all of the offspring from a single plant are members of a group called a clone. This simply means that their genetic make-up is identical to each other and to the lone parent. On the other hand, plants propagated by seeds, resulting from sexual reproduction, are not a clone because each seed has a unique genetic make-up, a mixture from two parents, different from either parent and different from one seed to another. The popularized term "cloning" conjures up science fiction images of laboratory witchcraft. But whether we call it "cloning," tissue culture, micropropagation, or growing *in vitro,* it is a significant horticultural propagation method which is experiencing rapid acceptance and will undoubtedly revolutionize the horticultural industry. (Fig. 1)

But growers still have many legitimate questions about the process. They want to know what tissue culture is, how it is done, who is doing it and why. They want to know if they can do it themselves, if they should do it and why. They are asking about the costs of building and operating a lab. They especially want to learn of the possible financial returns.

While tissue culture can serve a number of purposes the typical grower should consider tissue culture for two reasons: (1) mass production and (2) to establish and/or maintain "virus-free" stock. Other applications of tissue culture are less practical for the commercial grower for whom this book is written. Other uses include somatic hybridization (protoplast fusion), the induction and selection of mutants, and biosynthesis of secondary products. These intriguing applications will be discussed briefly in a later section to acquaint the reader with the state of the art and its impressive potential.

Tissue culture methods have been used extensively by the orchid industry for twenty years. Orchids have unique characteristics and problems; some early, fortuitous discoveries made it possible to tissue culture orchids of known quality where previously growers had struggled with unpredictable seed, or difficult to propagate, virus-infected stock. Very few growers of other plant families have done any tissue culture until recently, and rightfully so, because there have been too many unknowns, too many questions without answers. There will always be questions without answers, but enough valuable information has been gathered, and proven practical, for the average grower to reasonably consider tissue culture as a creative, promising, feasible option for propagating numerous plants, particularly those plants with special demands or difficulties. Only twenty years ago it was believed that woody plants could not be tissue cultured. It is now apparent that any plant can be tissue cultured if, or when, the right formula and processes have been developed for its culture.

When using conventional propagation methods one cutting produces one plant and one seed produces one seedling. In contrast, one explant (the culture start; i.e., one piece of stem, leaf, bud, root, etc. from the stock plant) theoretically can produce an infinite number of plants. Consequently very few stock plants are required to provide the explants needed to produce thousands of new plants. With the ever increasing value of land and plants the grower needs more than ever to put a dollar value on every square foot of ground and every stock plant. Because tissue culture requires a minimal amount of plant material to start with, significant savings can be realized by allocating fewer dollars, not only to the acquisition and maintenance of stock plants, but also to the space and time required to hold them.

Frequently a grower has a shortage of stock plants; only one new hybrid, for example, or only one mutant, one unique seedling, or just a few disease-free stock plants. Growers can produce commercial quantities in months by tissue culture instead of the years required by conventional propagation methods. In such a case the time advantage of tissue culture is very apparent.

Tissue culture provides the grower with several opportunities to realize savings of time and space. The rapid multiplication typical of tissue culture makes it possible to produce and sell more true-to-form plants sooner than by any other means. In the greenhouse, cuttings may take months to root. Tissue cultured plants usually take a fraction of that time; and while cultures are multiplying in the laboratory, greenhouse space can be used to finish off another crop.

Every year excessive amounts of grower time, labor, and room are spent on unproductive seeds, cuttings, and grafts. Significant numbers of young plants are lost to viruses, bacteria, fungi, insects, animals or other environmental factors. Some of these problems are derived directly from the stock plants from which the seeds or cuttings are taken. The diseases transmitted from parent to offspring often can be eliminated through tissue culture procedures. External contaminants such as bacteria, fungi, and insects are removed when cleaning the explant. Many internal viruses are eliminated by using the apical meristem as the explant. Apical meristem, the new, undifferentiated tissue at the apex, or very tip of a shoot, is usually virus-free in diseased plants because supposedly these cells grow faster than the viruses. If the few cells which make up the microscopic dome of apical meristem are removed from the plant and placed in culture they will grow and produce healthy, disease-free plants. This technique is known as meristem culture. It has been applied extensively both in research and commercially to eliminate a wide spectrum of viruses from lilies, dahlias, carnations, citrus, potatoes, and berries, particularly strawberries. Disease-free plants derived from tissue culture can retrieve much of the time now lost by growers because a higher percentage of clean, viable, mature, saleable plants will be produced.

Tissue cultured plantlets are not immune to attack and disaster but usually by the time they have come out of culture they are well started plants with a stockpile of nutrients, a good root system, and vigor unknown to cuttings. It is no secret that healthy plants are the first line of defense against disease.

Healthy plants are grown in the lab at any time of year. Most seeds and cuttings must be grown during a particular season, consequently work schedules in the nursery revolve around this factor. Tissue culture is not limited by the time of year or the weather. Working conditions in the lab are ideal and therefore conducive to year-round production scheduling, a situation that promotes maximum labor efficiency.

Tissue culture saves an enormous amount of the daily care required by cuttings and seedlings. Between transfers (periods of two to six weeks, after which cultures must be divided because they have multiplied) there is no need to water or tend the cultures other than casual surveillance. How different this is from the daily watering and weeding requirements accompanying greenhouse growing!

It goes without saying that profit is the ultimate purpose for a grower to have a tissue culture lab. Granted, there may be other satisfactions involved but, usually, profit is a primary reason for a commercial enterprise. The potential for profit in a tissue culture lab lies in producing a significantly greater number of healthier plants in less time, in less space, with less labor, and at less cost than by other means of vegetative propagation.

BOTANICAL BASIS FOR TISSUE CULTURE

The remarkable diversity of naturally occurring vegetative reproduction reflects the impressive potential of plants for multiplication. Their natural capability to reproduce themselves by asexual means is the basis for multiplication *in vitro,* or tissue culture. (*In vitro* means separate from the whole individual, but, very appropriately, the Latin translation means *in glass.*) The same multiplication and growth initiating factors which occur in tissue culture occur in the greenhouse and in nature. No new phenomena have been invented. Vegetative reproduction, whether occurring naturally or through human intervention, is initiated in stems, roots, or leaves. There is no mixing of gene traits which occurs in sexual reproduction within flowers.

We have taken many lessons from nature through careful observation of plant behavior and the study of plant physiology. From these lessons we have discovered some facts which enable us, as scientist and grower, to manipulate existing natural phenomena to serve our purposes. We take cuttings, we make divisions, we layer, we make grafts, and we use tissue culture. In short, we promote vegetative reproduction, a natural phenomenon. (Fig. 2)

Figure 1. A natural clone. Nature has been "cloning" for eons. Whenever a plant reproduces itself vegetatively it produces a clone.

Stems have tremendous potential for regeneration. They grow in many different forms and habits, long or short, slender or stout, above ground or underground, trailing or upright. Probably the vegetative propagation procedure most used by growers is that of growing new plants from stem cuttings. Stem cuttings are shoots or sections of stems that root when inserted into growing medium. Layering, another form of vegetative reproduction from stems, occurs frequently in nature and is widely used by horticulturists. A branch is said to layer when it comes in contact with the soil, roots, and grows a new plant. Wild blackberries rapidly expand their territory by layering. Layering is used commercially to propagate filbert, grape, black raspberry, trailing blackberry, currant, apple rootstock, and some ornamentals. (Fig. 2)

Growers also manipulate nature by grafting, attaching a piece of stem (scion) of one plant onto a rootstock of a different plant to obtain the desired traits of both.

Nature has other ways of multiplying from stems. True bulbs, such as lilies, tulips, and daffodils, are modified stems. They multiply naturally by growing bulblets in the axils of their scales. Growers encourage bulblet formation in some bulbs (narcissus and *Scilla*) by cutting them from top to bottom then further dividing them laterally. Each seg-

Figure 2. A grower-assisted clone. Growers have been "cloning" by cuttings for centuries. All of the members of a clone have identical genetic make-up.

ment will produce bulblets if it includes a piece of the basal plate. Growers promote bulblet formation in lilies by scaling. Bulb scales are removed from mature bulbs, dipped in rooting hormone and placed in growing conditions where they will produce bulblets. Scoring and scooping promote bulblet formation in hyacinth and *Scilla.* Scoring is merely making cuts across the basal plate. Scooping is to scoop out the basal plate. These methods simply induce the formation of more bulblets than would occur normally.

Corms are swollen underground stems complete with nodes, internodes and lateral buds; they do not have scales as do true bulbs. Corms, such as gladiolus and crocus, multiply naturally by producing cormels which are miniature corms.

Other modified stems that serve as vegetative reproductive structures are tubers and rhizomes. Irish potatoes, caladium, and gloxinia produce tubers, swollen underground stems. A piece of tuber or fleshy rhizome will produce a new plant if it contains an eye, or bud. Other rhizomes are long and slender, as in lily-of-the-valley and most perennial grasses. These rhizomes have long internodes with terminal and lateral buds that allow the plants to multiply effectively. Runners, as in strawberry plants, and stolons, as in *Cornus stolonifera,* are also modified stems. Their sole purpose is production of new plants.

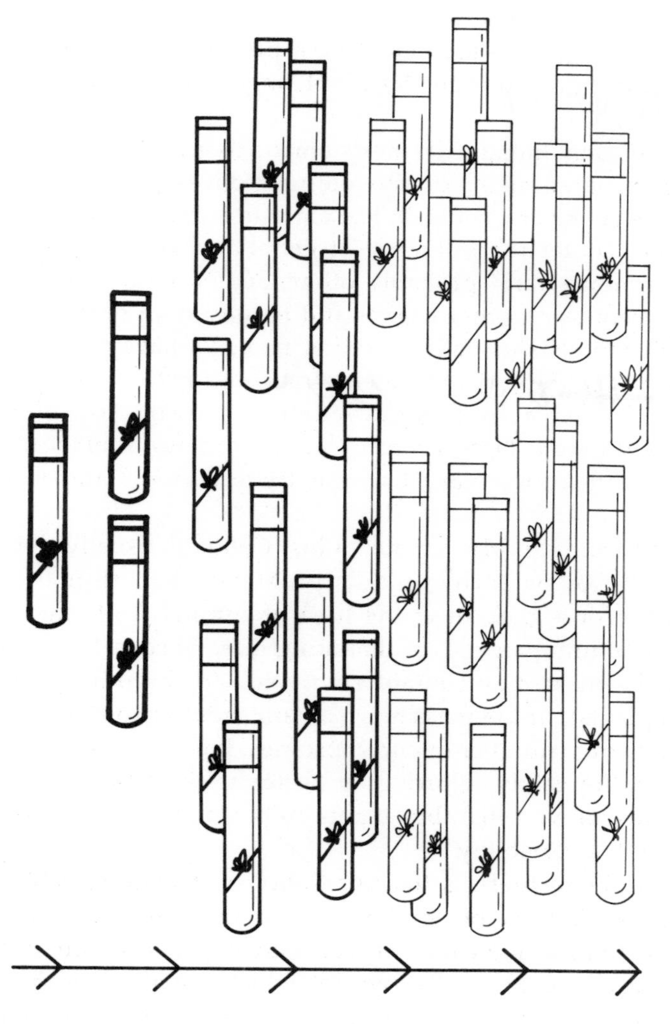

Figure 3. Tissue cultured clone. Potentially, any single explant can produce an infinite number of plants.

Roots take up water and nutrients from the soil and anchor plants firmly in the ground, but they also serve as vegetative reproductive structures, both in nature and commercially. Growers use root cuttings to propagate gooseberry, horseradish, apple, flowering quince, bayberry, aspen, and rose, to name a few. The "tubers" of sweet potatoes and dahlias are modified roots. The buds formed on the stem end of these tuberous roots take nourishment from the food stored in the tubers as they grow into new plants.

Leaves will occasionally produce new plants. Leaf cuttings are made from *Bryophyllum, Begonia rex, Sedum,* African violets, and *Sansevieria,* among others. In a few cases, new plants will grow from leaves without being separated from the parent plant; this happens in the piggyback plant (*Tolmeia*) and walking fern (*Camptosorus*).

The foregoing examples from field and greenhouse illustrate the inherent potential, the power, and the inclination that plants possess for vegetative multiplication. The multiplication of plants *in vitro* does not create new processes within the plants; tissue culture simply directs and assists the natural potential within the plant to put forth new growth and multiply in a highly efficient and predictable way. In contrast to sexual reproduction new plants produced through vegetative reproduction are actually severed extensions of the original plants. There is no parallel to this phenomenon in higher animal life. One cannot sever a toe and have it produce a new person. In short, plants possess a unique capability to regenerate not only tissues and organs but also whole functional plants. (Fig. 3)

New growth is usually initiated in meristematic tissue, tissue made up of cells which have not yet been programmed for their ultimate development. Meristematic cells are located at the tips of stems and roots, in leaf axils, in stems as cambium, on leaf margins, and in callus tissue. Under the influence of genetic make-up, location, light, temperature, nutrients, hormones, and probably other factors, meristematic cells differentiate into leaves, stems, roots, and other organs and tissues in an organized fashion.

Some differentiated cells (usually parenchyma cells) have the capability of reverting to a meristematic or dedifferentiated state to initiate growth of new and different tissue. Dedifferentiated cells often account for adventitious growth. Adventitious growth refers to the growth of new shoots, buds, roots, or leaves from unusual locations. Examples are aerial roots, buds from roots, plantlets from leaves, shoots and roots from callus of cuttings.

Dedifferentiated cells can also create callus. Callus is usually defined as a mass of undifferentiated cells, or parenchyma cells proliferating in response to wounding, as appears at the union of a graft, or at the base of cuttings. In tissue culture vocabulary callus is defined as an unorganized, proliferating mass of cells. Some of the cells may be differentiated, so the mass may contain embryoids (embryo-like structures capable of becoming normal plants), or the mass may contain shoot or root primordia, or there may be cells with an abnormal number of chromosomes. For example, many asparagus plants differentiating from the callus culture may be tetraploid (double the normal chromosome number in vegetative cells), but plants cultured from shoot tips, without a callus stage, do not show this variability.

Whenever cells divide there is the possibility of genetic variability. There are two kinds of plant cell division, meiotic and mitotic. Meiotic division relates only to sexually reproductive cells. Mitotic (somatic, or vegetative) cell division is the division of a vegetative cell to produce two cells each of which usually has the same number of chromosomes as the original cell. If a mutation—a change in chromosomes—occurs during mitosis, it is carried in all future divisions. Some somatic mutations go unnoticed because future cells are not affected; if a mutation for aberrant flowers occurs in a leaf

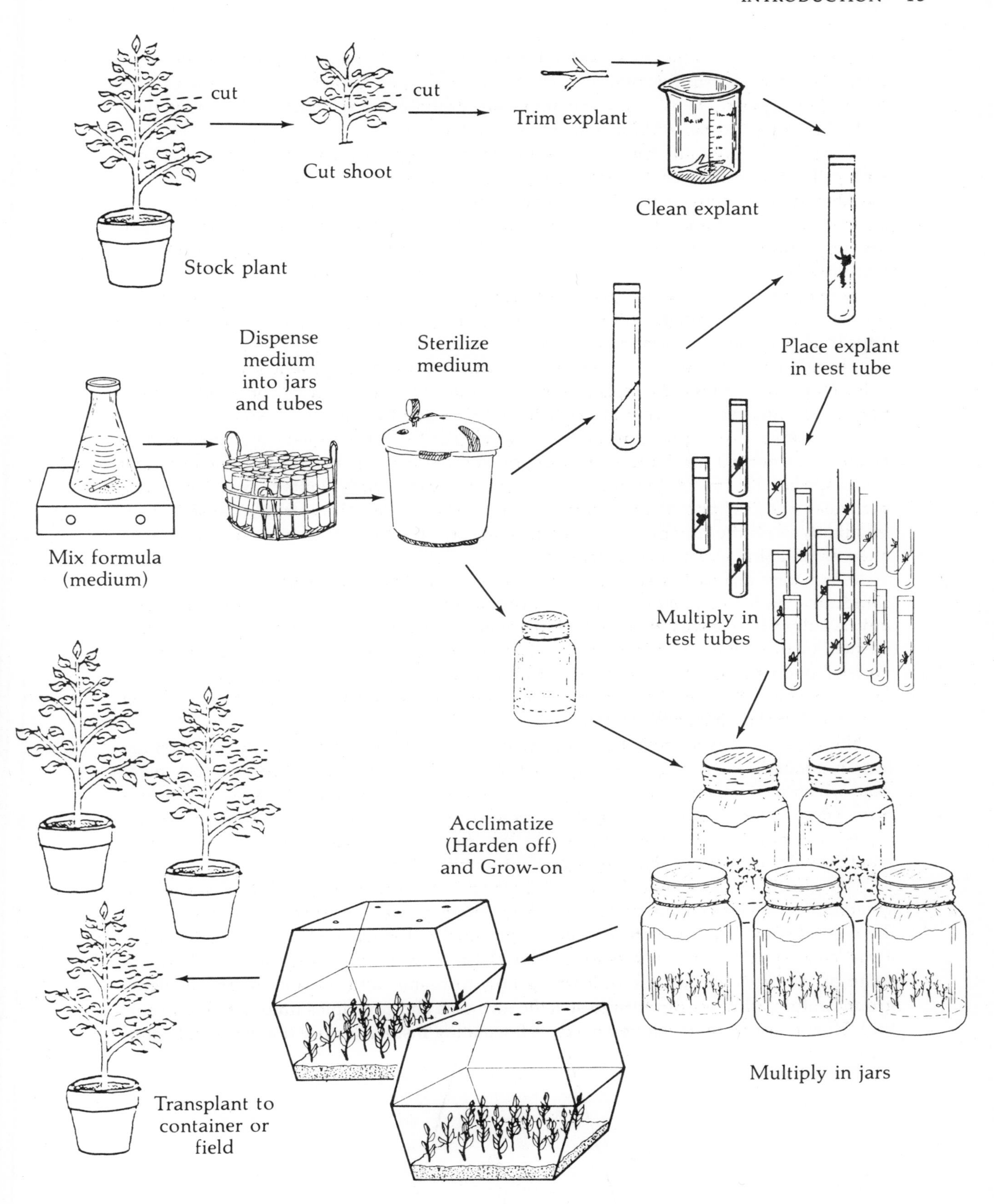

Figure 4. Sequence of shoot tip micropropagation.

bud, there will be no flowers to exhibit the mutation. Most mutations produce undesired effects: misshapen fruit or abnormal shoots, for example. Growers discard such plants when they appear. Occasionally mutations are desirable so scientists induce mutations with chemicals and radiation in a search for better plants.

Certain pelargoniums, *Sansevieria,* and other variegated plants are chimeras, plants with cells of different genetic make-up growing adjacent to each other. Some thornless blackberries are chimeras in which the aberrant characteristic (lack of thorns) is limited to epidermal tissue and can only be reproduced from stem tissue. Navel oranges and seedless grapes originated as sports and, obviously, can only be reproduced vegetatively.

Some abnormal divisions give rise to tetraploids. In the case of some giant apples, this is a desirable mutation; the opposite is true when huge, unfruitful grapevines appear, as they have in some vineyards.

There is concern among tissue culturists with respect to the genetic stability of tissue cultured plants. Thousands of plants could be cultured, only to reveal some defect when they are planted and growing in the field, a defect due to a mutation which was multiplied in culture. As a practical matter, however, cultures have been known to remain healthy and normal for decades showing no aberrations. It is helpful to remember that some species are more genetically stable than others. With limited experience in a particular species, one should start with numerous original explants, transfer them frequently, limit the number of subcultures, and start with new material annually or as often as on-going experience dictates.

When new material is started in culture it soon develops small shoots. If cultures could not revert to a juvenile state micropropagation would largely be impossible. The tiny, seedling size leaves which develop are appropriate to the size of test tubes while normal, mature leaves would rarely fit. The mechanisms of this return to juvenility are not well understood. After the tissue cultured plantlets are rooted and have grown in soil for a short time, they produce normal size leaves and assume the mature features of the plant stock from which they originated.

Not only do cultures revert to a juvenile state before they can multiply successfully but, in general, the younger the explant tissue, and the more juvenile the stock plant from which the explant is taken, the greater the chance the explant has to succeed in culture. The greater propensity of juvenile culture material to growth and multiplication suggests a familiar principle observed in the greenhouse, that cuttings from juvenile plants usually root better than cuttings from mature plants.

The principles of tissue culture are all around us, in nature, in the field, and in the greenhouse. We learn from experience, from other growers, or from reading. We learn the normal plant requirements for soil composition, nutrients, light, and temperature for a particular species. We study its form, growth habits, and how it reproduces. These are some of the clues we can use in learning how to tissue culture plants. With this background information we can turn to specific formulas and principles unique to tissue culture. How these were developed makes a fascinating history, a story that has just begun.

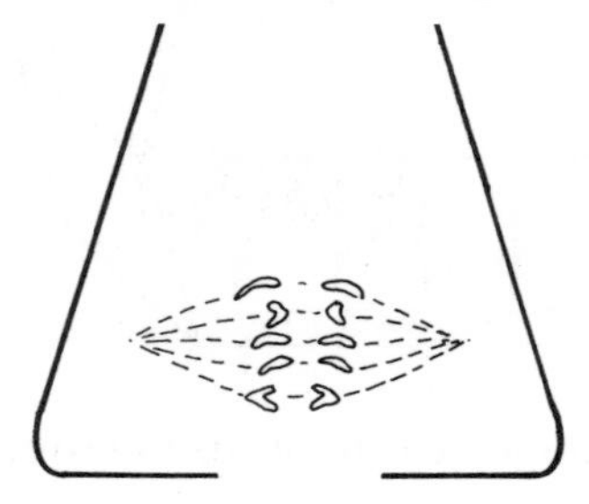

HISTORY

Tracing the rich history of tissue culture is as complex and challenging as tracking a genealogy. To some degree it involves the whole history of botany. However, a search through the years finds the unique contributions of certain individuals especially relevant; the events with which they were associated come together to provide us with an understanding and appreciation for the current state of the art.

One unique contributor was Zacharias Jansen, an eyeglass maker in Holland who, in 1590, invented the microscope. The invention and development of microscopes has permitted us to actually see cellular structure and accept as fact the role of cells as the building blocks of living tissues. Microscope lenses were later improved significantly by the Dutch merchant and politician, Antonj van Leeuwenhoek (1632–1723) who turned his glass blowing hobby to lens making. Marcello Malpighi, a seventeenth century Italian physiologist and pioneer in microscopic anatomy, alluded to cellular structure as tissue and compared it to woven cloth. In the same century Robert Hooke (a physician, mathematician, and inventor), of England, identified cells as nature's building blocks and was the first to call them "cells."

A botanist, M. V. Schleiden, and a zoologist, Theodore Schwann, further speculated on the nature of cells (1838–39). They observed that among lower plants, any cell can be detached from the plant and continue to grow. "We must, therefore," said Schwann, "ascribe an independent life to the cell as such." Such was the beginning of the theory of cell totipotence, the capability of any plant cell to regenerate the whole plant. Forty years later H. Vauchting tried, without success, to grow whole plants from pieces of plants. He was ahead of his time; he needed the tools of later discoveries.

Several other nineteenth century scientists deserve our attention. We must pay tribute to Charles Darwin, the king of observers, who first deduced the presence of a hormonal substance in grass coleoptiles. He observed them bending toward the light, yet, when he cut off the tips, the stems no longer bent. Another of the nineteenth century masters was Louis Pasteur who helped disprove the theory of spontaneous generation of organisms, and laid the groundwork for sterile technique. Justus van Liebig, a German physical chemist, theorized (1840) that the minerals in soil were essential to plant growth. His particular contribution was the concept of limiting factors: if one essential plant nutrient is missing, all the others present are of no benefit. Johan Knop (1817–1891) developed a nutrient solution based on his analysis of soil. "Knop's solution" was used by early investigators and is still useful today. By 1900 ten essential elements for plant nutrition were recognized.

In 1902 the German botanist, Gottlieb Haberlandt, predicted that plant embryos could be produced through the cultivation of vegetative cells. He realized the importance of cell culture as a tool to understanding the interrelationships of cells in the whole plant. Unfortunately his cultures were unsuccessful. However, E. Hannig, another German botanist, successfully cultured conifer embryos in 1904. He observed that premature, excised conifer embryos produced small, weak plantlets in culture instead of developing normal embryos. He called this "precocious germination."

As the twentieth century progressed, the field of plant tissue culture entered an era of exponential growth. Of early commercial significance was the germination and growth of orchid seedlings on agar medium in aseptic culture. This feat was independently accomplished, almost simultaneously, by L. Knudsen, Noel Bernard, and Burgeff in the early 1920's. About the same time W. Kotte (a student of Haberlandt) and W. J. Robbins independently cultured root tips.

Callus culture of carrot, a classical subject of investigation, was reported by two physicians, R. Blumenthal and P. Meyer, in 1924. They were primarily interested in its pathological implications. L. Reywald, however, demonstrated callus from carrot slices irrespective of pathological aspects (1927).

P. R. White is justifiably acknowledged as the father of tissue culture in America. He was the first to grow excised root tips in continuous culture. In 1939 he reported the successful culture of tobacco callus. In the same year, R. Gautheret and P. Nobecourt in France independently reported indefinite growth of callus from carrot cambium using auxin. In collaboration with Armin Braun, White demonstrated the similarity of plant and animal tumor cells by growing tumor tissue from tobacco. White's book, *A Handbook of Plant Tissue Culture* (1943) brought together the accumulated knowledge of plant tissue culture as it stood at that time. He also published on the value of glycine, pyridoxine and nicotinic acid in tomato root cultures.

Early attempts to tissue culture plants made it clear that plants in cultures were heterotrophic rather than autotrophic. Unlike plants in soil, plants in culture are unable to manufacture all of their own food, particularly proteins and carbohydrates, from inorganic nutrients. Perhaps empirically, it was discovered that sugar and undefined substances such as coconut milk, yeast, and fruit juices supported cultures where inorganic chemicals alone could not. Robbins reported in the 1920s that tomato root tip media was improved by the addition of yeast. Later analysis revealed that yeast contained desirable vitamins, particularly thiamin.

Inasmuch as some scientists were culturing embryos from seeds or attempting to stimulate embryogenesis (the spontaneous production of embryos from undifferentiated cells), it seemed logical to investigate the use of coconut milk, a ready-made, natural nutrient medium for embryos. Coconut milk was first used by J. van Overbeek and his associates in 1941. They discovered, with considerable excitement, that it stimulated callus formation in *Datura* cultures. F. C. Steward, a renowned plant physiologist at Cornell University, was so impressed by the dramatic effects of coconut milk in carrot culture media that he set aside his other objectives to study the growth factors in this and other liquid endosperms. Among the active materials extracted were several ingredients now commonly included in purified form in tissue culture media. Coconut milk, however, is still a common ingredient in orchid culture media.

The orchid industry was destined to be the first to apply micropropagation commercially. Based on techniques established by E. Ball (1946), George Morel and C. Martin cultured virus-free dahlia shoots and potato plants by meristem culture. In 1960 they applied their findings to orchids. They not only freed the orchids from virus, but implemented a process of multiplication that in time would change the features of the whole orchid industry.

Despite these successes, the need continued to determine defined ingredients for media and their proper portions for successful commercial application. Contributing to the discovery of necessary growth factors was Fritz Went. His work on plant hormones at the California Institute of Technology, together with K. V. Thimann, led to the discovery of the root promoting properties of auxins. In 1926 they rediscovered indole acetic acid which had first been isolated by E. Salkowski in 1885 but not previously recognized as a plant growth hormone. Went also collaborated with Folke Skoog to determine the bud inhibitory effect of auxin and its interaction with kinetin, the bud promoting cytokinin which Skoog and his associates had discovered at the University of Wisconsin. One important outcome of this work was an article by Skoog and F. C. Miller (1957), "Chemical Regulation of Growth and Organ Formation in Plant Tissues Cultured *in vitro.*"

Skoog's name is immediately recognized by anyone involved in tissue culture for his part in the universally used Murashige and Skoog medium formula. Commonly referred to as M & S or MS medium, it was reported in the now classic 1962 article, "A Revised Medium for Rapid Growth and Bioassays with Tobacco Tissue Cultures." Higher in salts than previous media, with subsequent modifications it has provided a magic key to culturing many more plants than previously had been possible. Elfriede Linsmaier and Skoog followed shortly with a report of their systematic study of the organic requirements of tobacco callus. Appropriate adjustments of the MS formula were indicated and the groundwork was laid for wider application of this versatile formula.

Toshio Murashige shares the honor of developing the valuable MS formula. Formerly a student of Skoog at the University of Wisconsin, and currently a professor at the University of California at Riverside, Murashige has done much to bridge the gap between research and industry. With untiring effort and almost single-handedly he introduced tissue culture to the commercial propagation world with contagious enthusiasm. His influence stretches still further through the works of his students, now professors, research scientists, and commercial tissue culturists in their own right. The consequent benefits, not just to propagators but to all citizens, exemplifies what should be the ultimate purpose of research.

New vistas in tissue culture research are continuously being opened in several overlapping areas. Among these are protoplast fusion, anther culture, mutagenesis, embryo culture, and the production of secondary products. Callus culture implies culture of a mass of unorganized, somewhat undifferentiated cells. Cell cultures are usually liquid suspensions derived from callus; however, cell suspensions of specific cells, such as leaf mesophyll cells, are also grown. Using growth regulators (hormones), and/or other factors, morphogenesis occurs. The term "morphogenesis" means initiation of form; therefore, for our purposes, the term translates to the formation of plant organs (shoots or roots) or somatic embryos (embryoids). When organs (shoot or root) are formed the process is called organogenesis; when embryoids are formed, embryogenesis. Cell culture has particular significance for mass production. With the potential for automatic introduction of fresh media into cell cultures thousands of embryoids may be initiated within a single container with a minimum of manual transfer. This is a dream of the future and the only way tissue culture will ever compete with the magnitude of nature's plentiful seed production. The fact that we can tissue culture clonal material of our choice (and possibly of our design) is, of course, the main advantage.

If the cell wall of a plant cell is dissolved away, the remaining membrane and its contents are called a protoplast. In general, because they are without cell walls, protoplasts are particularly useful for studying the uptake of, or resistance to, toxins, nuclear material (DNA), virus, bacteria, or fungal cells. Of particular interest to plant breeders is the fact that protoplasts will fuse with other protoplasts, not only of the same clone but with those of other species or genera. Before 1960 walls were removed by physical means (micro surgery). First, the cells were plasmolyzed by placing them in a solution which caused the protoplasts to shrink away from the walls. Next, the cells were randomly cut up with a sharp scalpel. The undamaged protoplasts were recovered with a pipet. In 1960, E. C. Cocking, in England, published a method for removing cell walls by chemical (enzymatic) methods.

Complete plants have been regenerated from protoplast cultures in at least 30 species including tobacco, wheat, carrot, asparagus, potato and tomato. After many failures by numerous investigators to induce fusion between protoplasts from different sources Kao and Michayluk, in 1974, published "A Method for High Frequency Intergeneric Fusion of Plant Protoplasts." They discovered that when a solution of polyethylene glycol was added to a mixed protoplast culture followed by high calcium dilution with high pH that a relatively high percentage of the protoplasts fused. Another inducing agent discovered more recently is that of an electrical impulse. Protoplasts of tomato have been fused with protoplasts of potato and differentiated into complete hybrid plants. The objective was not to create a freak but to convey certain disease-resistant qualities of one plant to another by means of an intermediary plant. So far, only very few fusions between different species have developed into mature plants, but the potential is there and the technology is evolving.

Interest in anther culture is spurred by the possible practical applications of haploid cultures (cells having only half the normal number of chromosomes in vegetative cells, but the normal number for sexual cells). Anthers are removed from young flower buds after meiosis (reduction division) but before the pollen is fully developed. Haploid plants grown from anther culture have sterile flowers, but normal plants develop if the chromosomes spontaneously double or are induced by chemicals to do so. Doubling can also be achieved by protoplast fusion. The seeds from such diploid plants are true to type, so it is little wonder plant breeders are interested in anther culture, considering it takes several generations of inbreeding by conventional means to obtain a pure line.

Mutagenesis is the production of mutants, whether spontaneously or induced by artificial means. Tissue culture offers a convenient method of handling plant material when it is subjected to mutagens (mutant inducing agents) and subsequent screening. Treatment may be by radiation, ultraviolet light, or certain carcinogenic chemicals. Following treatment, cultures are incubated and tested for characteristics such as resistance to toxins, salts, herbicides, antibiotics, or disease. These treated cultures are also tested for their tolerance for heat or cold, hormone and nutrient requirements, or their production of secondary products.

Many hybrid plants produce viable embryos that do not mature to viable seeds. Embryo culture is providing a solution to this problem. Not only does it rescue valuable embryos, but also multiplies them. *In vitro* fertilization and subsequent embryo culture is a new vital method for interspecific and intergeneric hybridizing that is also receiving wide interest.

In a different vein, the production of secondary products from tissue cultures is a highly promising application. Personnel from oil and food producing companies are prominent among students of tissue culture. It is conceivable that cell cultures can be coaxed to produce greater quantities of useful compounds, under controlled conditions, with maximum efficiency, leading to greater availability at less cost, than from whole plants. A sampling of the types of substances potentially available from cell cultures includes flavors, pigments, medicinals, antibiotics, insecticides, fungicides, alkaloids, oils, and latex.

This brief coverage of some of the current research activities involving tissue culture should be enough to stimulate the imagination and to alert the commercial propagator to the broad potentials and promise of these procedures. The tangible results of these research efforts will come as no surprise to the knowledgeable propagator. They will appear one by one in the nursery trade, in the sweeping fields of grain, in the kitchen, in the flower shop, in the now empty bowls of the hungry, and in everyday technology.

References: #49, #50, #64, #72, #109, #113, #114, #133, #141, #163, #164.

Chapter II: The Laboratory

FACILITIES

Basic requirements

No matter what size of laboratory is decided upon, three distinct lab areas must be established. These are a preparation area, a transfer chamber, and a room in which to grow the cultures. In extreme or temporary circumstances these three areas might occupy the same room, but long term operating efficiency and the need for cleanliness dictate each be in a separate room.

The preparation area resembles a kitchen. There should be a sink with hot and cold running water, a refrigerator, and a dishwasher. Provision must be made for preparing media, the nutrient-containing liquid, or agar-solidified base, in or on which the cultures grow. Media must be heated and stirred, dispensed, sterilized and stored.

The transfer chamber is a sterile box, or hood, where the technician starts, divides, trims, and transfers the cultures from one container to others. (Figure 8-2)

Well lighted shelves in a warm, clean room are needed to hold the tubed or bottled cultures while they are growing.

Location

From the planning stage on, careful provision for a clean environment is required if the tissue culture program is to be successful. If the lab is to be a new, free-standing structure, it should be placed where access will be paved and the dust and chemicals of the field will not blow on it. Smoke or other neighborhood air pollutants should also be avoided.

It may be desirable to plan an inexpensive, temporary set-up to prove that the practice of tissue culture is appropriate, feasible and applicable to the requirements of the nursery. The trial lab could be a remodeled existing structure—a house, an office, a warehouse, a greenhouse, or other building, or part of one of these. There are advantages to a free standing lab; it will have good access and no contamination from adjacent rooms; however, an addition to an existing structure may cost less and provide more convenient access from other operations. In any case, cleanliness must be the first guiding principle.

A hobbyist may start tissue culturing in a home kitchen, using a homemade transfer chamber (Figure 8-3) and a bookcase with lighted shelves. These will serve quite well for awhile, but the serious amateur who masters sterile technique and achieves rewarding results in limited facilities will want to advance to a more convenient laboratory for pleasure and profit.

Design

Cleanliness is not only important in the siting of a lab, but the major consideration in its design. The cleanest prospective work flow and traffic pattern should govern the planning of room location, passageways, work areas, doors, and pass-through windows. The transfer room and the culture growing room should be isolated as much as possible from outside doors and major foot traffic. Driveways and parking areas should be close enough for convenience but far enough away that blowing dust will not affect the lab. Well-kept lawn and disease-free shrubs will not only contribute to an attractive entrance

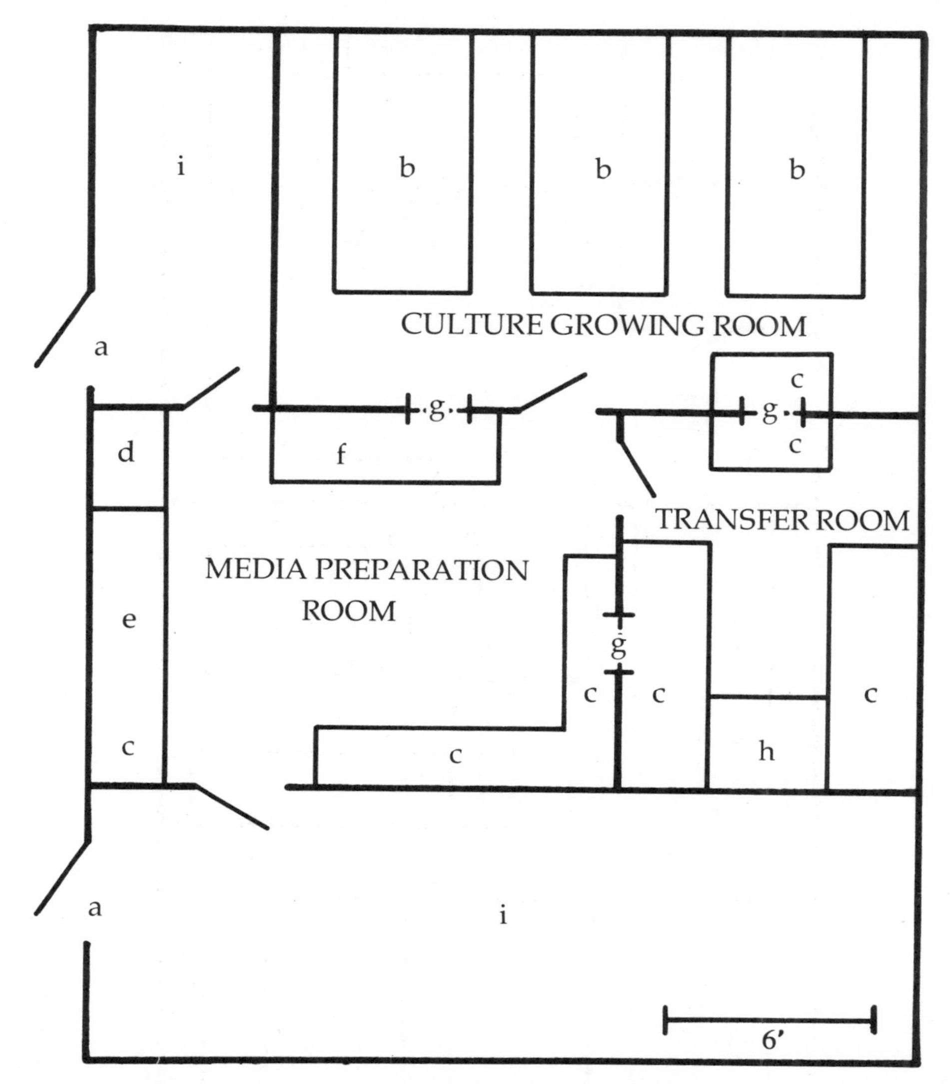

Figure 5. A small tissue culture lab (author's original lab). (a) Entry. (b) Shelves. (c) Counter. (d) Refrigerator. (e) Sink. (f) Stove top burner. (g) Pass-through window. (h) Transfer chamber. (i) Expansion. See text.

but promote a clean atmosphere as well.

A simple plan for an operation incorporating some of these design features and housing two technicians is shown in Figure 5. The media preparation room is 16′ × 12′, the transfer room is 8′ × 12′, and the culture growing room is 18′ × 12′. The plan lends itself to various options. The transfer chamber can be widened to allow two technicians to transfer at the same time. Either entry can be a simple passageway or be developed further for storage, office, autoclave, water treatment, or rest rooms. In this plan there are no upper cupboards because below bench space provides adequate storage space and better bench access is available without overhanging cabinets. Windows, if desired, may be placed wherever convenient in the media preparation and transfer rooms. There should be no windows to the outdoors in the culture growing room because outdoor light and temperatures reduce the ability to control the growing environment.

Other lab designs are shown in Figures 6 and 7.

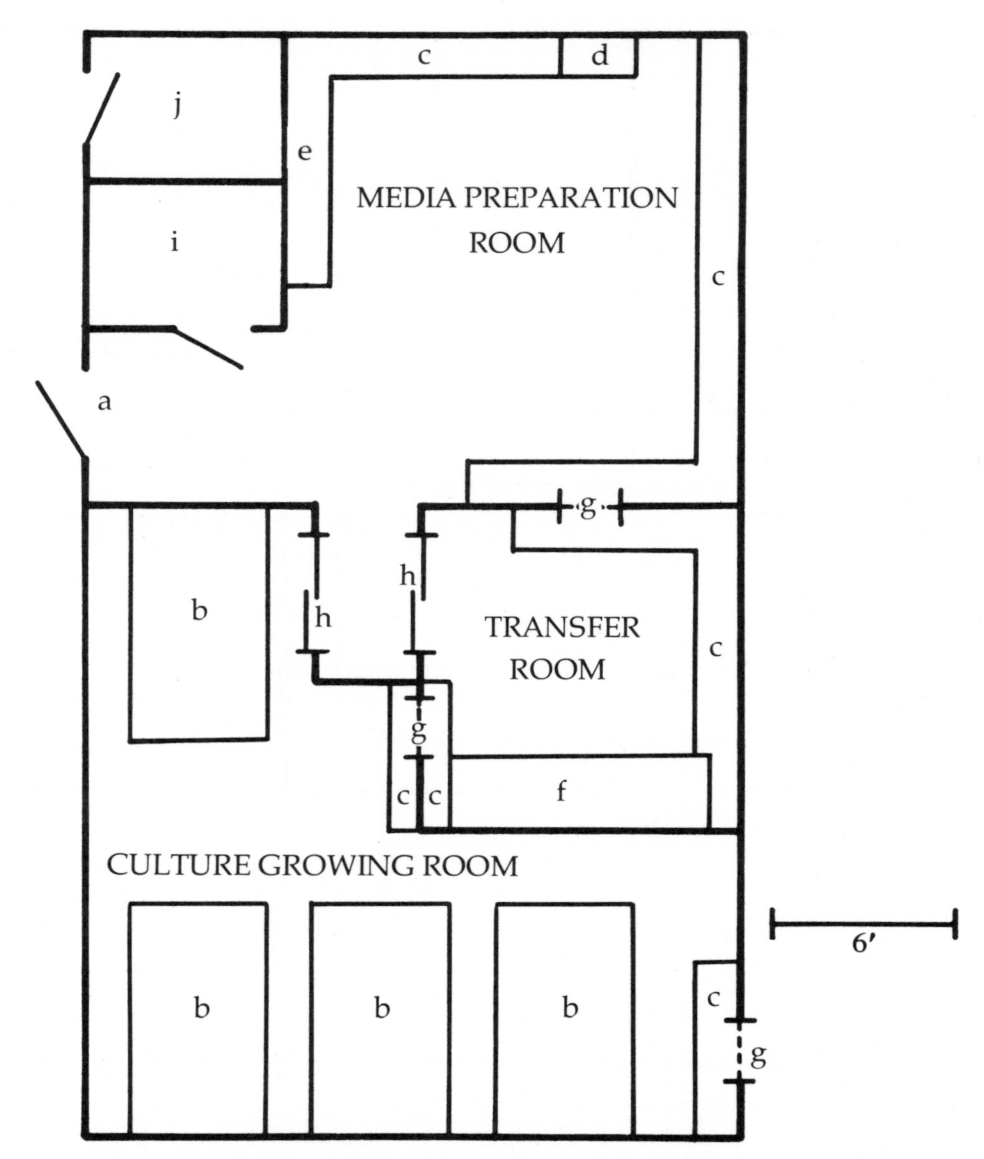

Figure 6. A medium-sized tissue culture lab. (a) Entry. (b) Shelves. (c) Counter. (d) Refrigerator. (e) Sink. (f) Transfer chamber. (g) Pass-through window. (h) Sliding glass doors. (i) Storage. (j) Restroom.

Figure 6 is similar to Figure 5 but is for a somewhat larger operation. The preparation room (Figure 6) is 15′ × 15′, the transfer room is 10′ × 10′, and the culture growing room is L-shaped with 312 sq. ft. The entry allows access to transfer and growing rooms with minimal traffic through the preparation room. Windows looking outside line one wall of the preparation and transfer rooms. Sliding glass doors provide access to and immediate visibility of the transfer and growing rooms.

Figure 7 is a design for a much larger operation than the other two plans. It can easily accommodate six technicians. The three rooms—preparation room, transfer room, and growing room—are individually accessible from an outer hall or walkway. The preparation room is supplemented with a storage room, a rest room, and a service room in which the water heater, water treatment equipment and an autoclave are located. Access

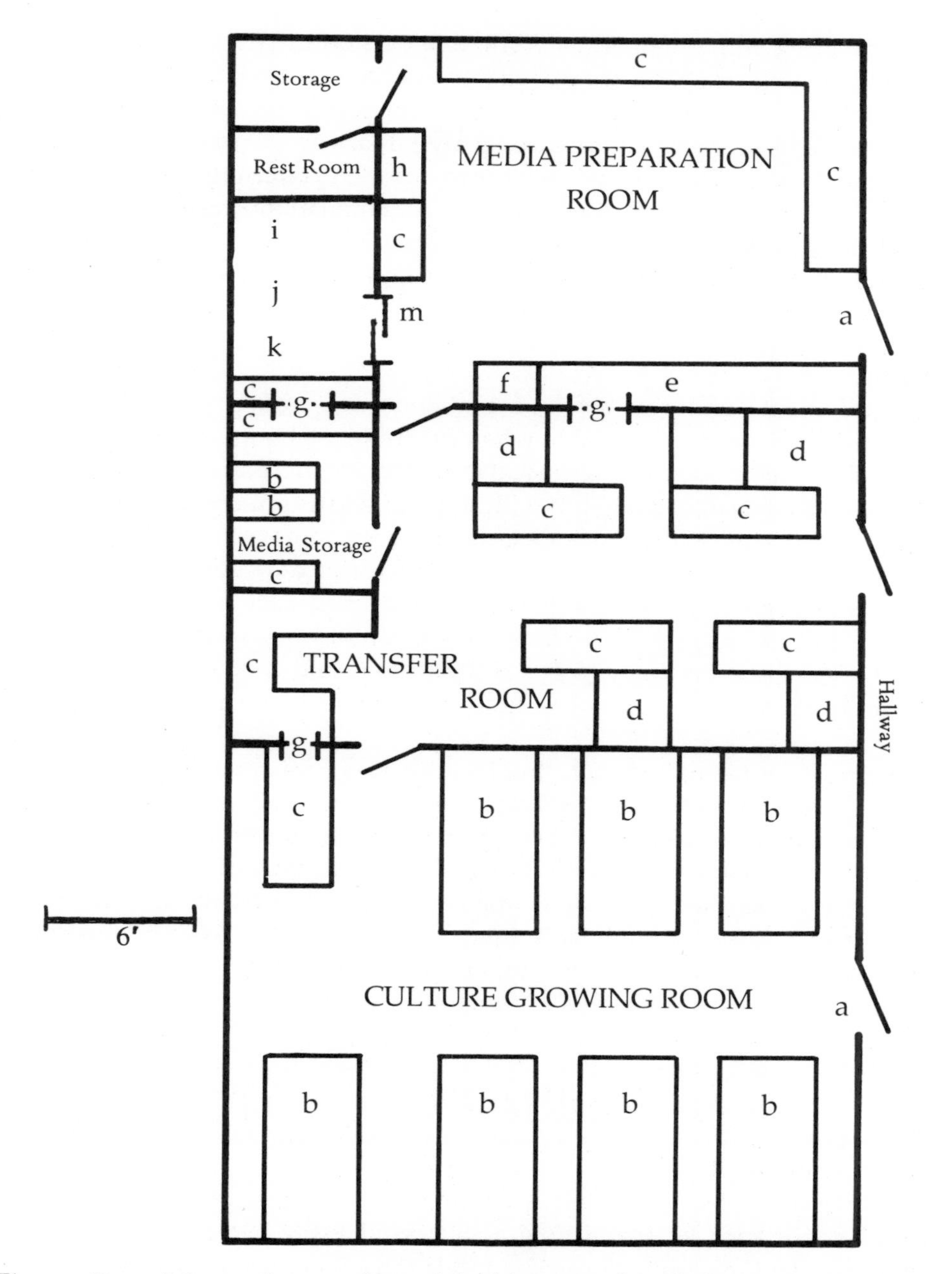

Figure 7. A large tissue culture lab. (a) Entry. (b) Shelves. (c) Counter. (d) Transfer chamber. (e) Sink. (f) Dishwasher. (g) Pass-through window. (h) Refrigerator. (i) Water heater. (j) Water treatment. (k) Autoclave. (l) Sliding glass doors.

to the service room should be provided by sliding glass doors for easy monitoring of equipment and wide access to materials being moved to and from the autoclave. In the preparation room available counter space can be increased by a bench or work table in the middle of the room. All aisles and doorways should be wide enough to accommodate carts. The size of cart can vary from tea cart to bank of shelves on casters.

The transfer room in Figure 7 is designed for four technicians. In the version illustrated the technicians have permanent side counter space on which to place cultures and materials, thus providing ample individual work area outside the transfer chamber. An alternate plan has two six-foot chambers placed back-to-back in the middle of the room. Common walls and wiring in this alternate arrangement provide a construction advantage. In such a case the technicians' cultures and materials are on carts alongside them.

In all three model plans pass-through windows between rooms are recommended. They save time by reducing traffic, they reduce contamination by limiting airflow through opened doors, and they allow good visibility. Other construction features in common include 4′ × 8′ wall paneling, suspended ceiling panels with recessed fluorescent lights, and vinyl floor covering.

Power and wiring

Electrical requirements and fire safety precautions are sufficiently important to deserve the best professional help available when planning a laboratory. This statement is true of any building plan, but it is especially true when there are several unusual requirements, as in a tissue culture lab. If the fluorescent fixtures in the culture growing room give off too much heat, then the ballasts must be removed and placed on a protected panel outside of the lab. Some of the newer fixtures now being manufactured do not present this heat problem because the old style ballasts have been replaced with smaller, solid-state ballasts, or miniature transformers. Culture growing room temperatures are usually maintained between 75°F and 85°F (24°C and 29°C) with 16 hours of light and 8 hours of darkness. Twenty-four hour timers will automatically control the lighting.

Most of the wiring can be for 110 volts, but heating, water treatment equipment, and autoclave may require 220 volts. Heating and cooling for the lab can be supplied by electric wall heaters and air conditioners, but a heat pump system is far better and well worth the initial expense. Some means of filtering lab air or air intake is highly desirable. If a heat pump system is installed it should contain an in-line electrostatic air filter. Rapid new developments in heating, lighting, and energy conservation make it imperative to confer with your power company and heating and lighting manufacturers' representatives before making final decisions in this important aspect of lab planning.

EQUIPMENT

Water purification equipment

Water is the largest component of tissue culture media so the quality of the water used is of critical importance in establishing a successful tissue culture lab. Tap water usually contains dissolved minerals. These chemicals must be removed before using the water in tissue culture media because they upset the precise balance of nutrients in media formulas. This upset can promote precipitates (insoluble compounds) that make the nutrients unavailable to the plantlets. In addition, by introducing excessive amounts of unknown chemicals a nutrient imbalance or a toxic medium may well result.

Water purity is usually measured by its resistance to electrical current—the greater the resistivity the higher the purity. Resistivity is measured in ohms-cm. Water for commercial tissue culture should have a resistivity of at least 200,000 ohms-cm. Some meters indicate the reciprocal of resistivity which is the conductivity of the water. Conductivity is measured in mhos/cm. Water with a resistivity of 200,000 ohms-cm has a conductivity of 5.0 micromhos/cm (.000005 mhos/cm). An inexpensive conductivity meter costs about $400. Most companies dealing in purified water and purification equipment will test water samples without charge.

The three common methods used to remove dissolved chemicals from water are distillation, deionization and reverse osmosis. Sometimes a combination of two or more methods is required.

Distilled water has been the standard laboratory-grade water for many years. Water is boiled in a still leaving nonvolatile salts in the boiler. The steam is condensed to water in the condensing coil and collected in a container, ready to use. A free standing, home use, electric stainless steel still (Durastill) is available for as little as $300. It occupies about two square feet of bench space and produces a gallon of distilled water in three hours using about 3 kilowatt hours of electricity. Purity of the end product depends on the feed water and still efficiency. At five cents per kilowatt hour the direct cost of one gallon of distilled water from this still is about fifteen cents.

Deionization is another effective way to remove dissolved chemicals from water. A single mixed resin bed tank, or cartridge, will often provide adequate purity for normal tissue culture production. This system is connected to existing water lines and provides continuous flow. In this process water is passed through a mixture of positive and negative ion exchange resins. The undesirable ions in the water are exchanged for H^+ and OH^- ions on the resins. These H^+ and OH^- ions combine to form water while the undesirable ions replace them on the resins. If the feed water is relatively free of impurities the resins work efficiently for several months. However, when the feed water is "hard" and/or contains a high level of impurities, the exchange resins must be regenerated, or new cartridges purchased, more frequently. Resin bed exchangers are usually equipped with a warning light to indicate when the resins are spent. The light is set to indicate when the water conductivity is becoming too high. The higher the conductivity, the greater the degree of saturation of the resins with undesirable ions. Companies which sell pure water and water purification equipment both rent and sell mixed resin tanks which they will regenerate for $100, and up, depending on the size. They also sell disposable cartridges.

Various systems using disposable cartridges are available from scientific supply companies. One such system offers a demineralizer (Barnstead's Bantam D0800B) which includes a stand to hold the cartridges, a conductivity meter with indicator, and connecting hose, all for about $389. Among disposable cartridges to fit this unit is one for about $46 (Barnstead D0809). This unit can remove a total of 104 grams of calcium carbonate. For example, if the feed water has a resistivity of 5000 ohms-cm because it contains 81 mg of calcium carbonate per liter, this unit could purify 1284 liters of feed water. If lab use is 10 liters a day the cartridge will have to be replaced in five months. At a cost of $45 for the cartridge, the direct cost of one liter of purified water in this case is about three and a half cents, exclusive of the original investment.

The third method of water purification is reverse osmosis which is usually used in combination with a deionizer or still. It does not have the refinement of the other two methods but provides excellent pretreatment. Osmosis is the diffusion of a fluid through a membrane into a solution of higher concentration to equalize concentrations on both sides of the membrane. Reverse osmosis forces the solution through a membrane in the opposite direction filtering out the impurities. Raw feed water is run through the reverse osmosis system first to remove most of the impurities before flowing through a deionizer or entering a still. The efficiency of the subsequent still or deionizer is increased significantly. Reverse osmosis systems are initially expensive but combined with a still or deionizer, and given the right feed water (nonacid water, low in calcium and iron) high grade water (.06 micromho) can be obtained.

A combination system (Barnstead) that delivers 10 liters per hour costs about $5000. Reverse osmosis membrane replacements for this system cost about $400 but may last up to two years depending on the feed water. By using reverse osmosis pretreatment manufacturers claim to reduce deionizing cartridge replacement costs by a factor of eight. A similar system (Continental Water Systems Corporation) claims a cost of about 30 cents per gallon of .06 micromho/cm water.

Both deionized and distilled water are available at local supermarkets in gallon jugs at less than a dollar a gallon. It is advisable to buy treated water when starting on a small scale and limited budget. Rain water may be used, but check its conductivity. Water with as much as 15 micro-mhos/cm conductivity is acceptable though not ideal.

Balances

Unless premixed chemicals are to be used, a precision balance (scale) will be required to accurately measure the small amounts of chemicals required for tissue culture media. Most analytical balances cost over $1900. These expensive, electronic balances are fast and precise, a worthwhile investment if affordable. There are several kinds of industrial or pharmaceutical two-pan balances available (dial weighing, hinged cover) with 2–5 milligram sensitivity which, while slower and less precise than analytical balances, are satisfactory for weighing 10 milligram quantities or more. These balances cost about $500 and are available from most scientific supply companies. For weighing over 1 gram they require a set of weights (1 gram to 50 grams) costing an additional $50 to $300 depending upon the number of weights, accuracy, and construction.

Even if the capacity of a balance with 2 milligram sensitivity is over 100 grams it is desirable to have a triple beam balance for weighing over 50 grams. A triple beam balance is usually faster and simpler to use for weighing quantities over 10 grams and using it instead of the analytical balance preserves the more sensitive instrument for weighing minute quantities. Triple beam balances cost about $100.

pH meters

The acidity or alkalinity (pH) of tissue culture media is important and specific to plant requirements just as it is in soils and potting mixes. For example, rhododendrons are tissue cultured in a relatively acidic medium (pH 4.5–5.0), while strawberries require a less acidic (pH 5.7) medium.

While a pH meter is necessary for commercial production, a beginner can get by with pH indicator paper. Recommended pH ranges for test papers are 2.9–5.2, 4.9–6.9, and 5.5–8.0. These small ranges will measure adjustments of media to approximate pH requirements.

Every commercial lab should invest in a pH meter. A pH meter costs from $100 up. A good quality meter costs in the neighborhood of $600 and is a wise investment.

Hot plate/stirrer

Agar media tend to stick and burn when heated unless they are constantly and effectively stirred until they boil. A combination hot plate/automatic stirrer is among the most useful aids in preparing media. A rotating magnet built into the hot plate/stirrer causes a magnetic stir bar, placed inside the flask of medium (Fig. 9. c), to rotate and so prevent sticking. The stirrer and hot plate features can be used at the same time or independently. A hand held stirring rod, beater, or spoon will also work but with a greater likelihood of the agar sticking (unless the solution is in a double boiler). The hot plate/stirrer will not work with most metal containers because the stir bar is magnetic. The stir bar is usually used in a glass container. An additional, very important use for the automatic stirrer is to provide agitation when cleaning explants. (See explant cleaning equipment section below.) A hot plate/stirrer costs about $275 and up.

A gas burner will heat much faster than a hot plate making it more desirable for larger quantities of media. In this case you will need a large cooking vessel, possibly a canner, and a top mounted stirring motor and stirring rod ($300).

Explant cleaning equipment

Some laboratories use a vacuum pump routinely to help disinfect explants. The suction of the vacuum (25 mm of mercury) improves contact between the cleaning solution and the explant and helps to disengage contaminants. The need for a vacuum pump, or aspirator pump, should be established before investing in one, because they are cumbersome to use and too high a vacuum can injure the explants. If a lab experiences continuing problems with contamination of new material a vacuum system is a viable option.

An ultrasonic cleaner is another option to help remove contaminants. The action of the ultrasonic cleaner causes different materials to vibrate at different rates thus dissociating them. The contaminants are literally shaken loose from the plant material. A dental lab ultrasonic cleaner costs about $100.

A magnetic stirrer (hot plate/stirrer, see above) usually provides adequate agitation in the cleaning of explants.

Media dispenser

An automatic pipetter to dispense media into test tubes or jars is a valuable labor saving device for a production lab. While a good technician with a quick eye and a steady hand can pour 100 tubes in 10 minutes from a liter pitcher, the automatic dispenser will fill 500 tubes in the same time.

A small lab can easily postpone purchase of an automatic pipetter. The innovative technician will dispense media from a coffee urn, or establish a reservoir for media that is gravity fed through tubing with a pinch clamp.

An automatic pipetter costs about $1000.

Sterilizing equipment

A modest lab operated by no more than three people can readily use a household pressure canner (cooker) for sterilizing media in tubes or jars. A larger lab should purchase an autoclave. Because a pressure cooker is round it does not efficiently accommodate standard, rectangular holders. As a consequence test tubes and jars must be handled more frequently when using a cooker which becomes inefficient in a large lab. Autoclaves are automated, designed to handle retangular holders, and are larger than pressure cookers. The cost of an autoclave ranges from $2000 to $12,000, depending on the size. The cost of a pressure cooker is about $75. If a pressure cooker is used for sterilizing, a stove or stove top unit is necessary for heating the cooker.

In using a standard pressure cooker (canner) to sterilize test tubes containing media it is necessary to buy or construct a wire basket similar to the baskets which used to accompany the cookers but are no longer available. A wire frame with handles is lined with ½″ or ¼″ hardware cloth to prevent the test tubes from falling through. A basket can be constructed as illustrated in Figure 11-a.

Rotator

Some cultures require liquid media for optimum performance. Such cultures should be gently agitated by means of a rotator (1 rpm). Agitation aerates the medium thereby preventing the culture from "drowning." Agitation also disperses the waste products of the culture. A rotator and head holding 108 16 mm test tubes costs about $500. One can be built for less than $100 and a little ingenuity.

Dishwasher

A conventional, built-in household dishwasher will handle the dishwashing requirements of a small laboratory (1000 test tubes per day, or less). If the prongs on the bottom rack are bent down, 160 test tubes in test tube racks (40 to a rack, Figure 10-a), can be washed at one time. A commercial laboratory dishwasher of the same capacity will cost at least four times as much as a kitchen dishwasher and yield few advantages. A large commercial washer would be considered in a larger laboratory where the operating tube level is over 1000 tubes a day. A household dishwasher costs about $400. A laboratory glassware washer costs from $2000 to $7500.

Refrigerator

A household refrigerator is useful to store perishable chemicals and stock solutions. A larger refrigerator is necessary if media or cultures are to be refrigerated. A walk-in refrigerator (1°–3°C) is a valuable means of delaying growth of cultures (1) when workload prohibits immediate attention, (2) when the plantlets are ready too soon to go into the greenhouse, or (3) for holding stock cultures. An expansion coil with fan, freon compressor, and temperature controller are the basic "off the shelf" components sized to your needs. We use a 24" × 24" exchanger (expansion coil) and a ¾ horsepower compressor for our 10′ × 10′ × 7′ high refrigerated storage room.

Labeler

Different varieties of plants are not easily distinguished in culture so it is extremely important to identify each container at the time it is processed. A practical labeling procedure is necessary in even the smallest tissue culture lab. Every system has its drawbacks. While marking pens will work, they are slow and any moisture on glass or pen makes the pen skip. If one uses a marking pen with permanent, waterproof ink, the ink must be scrubbed off with a pot scrubber. If one uses a marking pen with ink that washes off (overhead projector marker) the ink may rub off prematurely. Hand stamps are also a problem due to permanent ink. Both markers and hand stamps are too slow for commercial production. Grocery store price labelers are proving satisfactory for many labs. The least expensive kind costs about $100. If the labels come off in the dishwasher they may clog it; however, if they are on lightly washed lids the labels can pile up for easy, infrequent removal.

Microscope

Whether a lab should purchase a microscope or not depends upon the type of tissue culture work to be done. A dissecting microscope is required for obtaining meristem starts. A compound microscope is not necessary for routine, commercial production but is required by labs doing sophisticated work.

Even when a microscope is not essential in production (when the explants are not microscopic), it is a helpful instrument to have available. It is an extension of our limited vision and a challenge to the curious mind.

A $500 dissecting microscope is satisfactory.

Transfer chamber (hood)

Laminar air flow transfer hoods are available commercially. They provide a sterile atmosphere in which to work with the cultures. Air is forced through a HEPA (high effi-

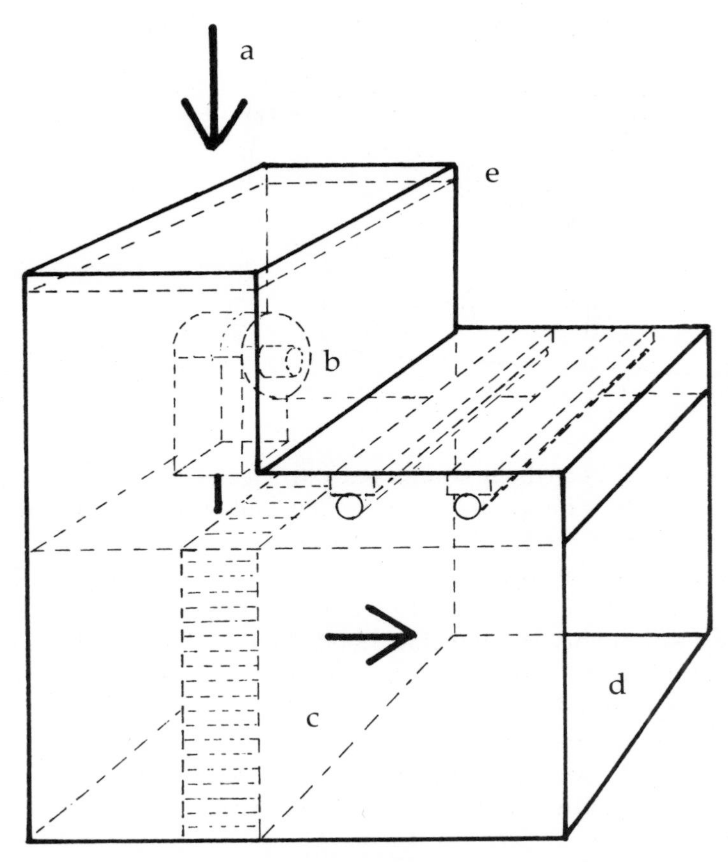

Figure 8-1. A laminar air flow transfer chamber showing (a) airflow, (b) blower, (c) HEPA filter, (d) work area, and (e) replaceable prefilter (furnace filter).

Figure 8-2. Detail of work area.

Figure 8-3.
A still-air box for the hobbyist.

ciency particulate air) filter which strains out particles as small as .3 micron assuring a high degree of sterility. The gentle, scarcely detectable air stream flows through the HEPA filter at the rear of the hood toward the technician (at 100 fpm across the workbench) providing a sterile atmosphere in which the technician works. The filters can be purchased separately (2′ × 3′ for $250) so that it is possible and practical to construct rather than buy a satisfactory laminar flow hood (Figures 8-1,2). Additional components are a prefilter (inexpensive furnace filter), a blower, plywood sides and top, a smooth bench, and a fluorescent light.

Laminar flow hoods are essential for commercial operations. A still air transfer chamber with a slanted glass front and partially enclosed hand access is a viable alternative for the hobbyist (Figure 8-3). Some hobbyists have even worked within large, clear plastic bags.

A laminar flow hood to accommodate one technician can be built for less than $600. A commercial one costs about $1500.

Growing room shelves

Lighted shelves must be constructed to hold the tubes or jars of cultures while they are growing in the culture growing room. One way to build them is to use slotted steel angle as supports for 4′ × 8′ shelving of particleboard, plywood, or wire mesh (expanded metal, or ¼″ or ½″ hardware cloth). Wire mesh is more expensive but allows better air circulation than solid shelving. Boards should be painted white to maximize available light. A room 8′ high will accommodate 5 shelves 18″ apart with the bottom shelf 4″ off the floor. Purchase a light meter reading in foot candles to measure light distribution and intensity. A small light meter costs about $60.

Two 8′ cool white fluorescent fixtures spaced 20″ apart supply 100 to 300 foot candles of light to the cultures on each 4′ × 8′ shelf. A 24-hour timer automatically turns the lights on and off.

Recent developments in fluorescent fixtures modify or eliminate the problem of heat given off by ballasts. Hot spots in the shelves directly over the ballasts have often experienced temperatures sufficiently high to cook the cultures; therefore, in some labs the old style magnetic ballasts have been removed and placed at the end of the shelves or installed outside the room. The newer fixtures, however, have either a smaller, electronic ballast, or a very small transformer in one end of the fixture. Some of these new, solid state ballasts offer advantages of less heat, no flicker, and no noise. Consult your local lighting or electrical authorities for technical information.

One rack consisting of five 4′ × 8′ particleboard shelves, metal framing, and lights costs about $500.

Use and care of equipment

Most laboratory equipment is quite different from the equipment used by growers in nursery operations. In particular, lab instruments are designed to make precise measurements. As precision instruments they are both delicate and demanding of very careful use—unlike tractors or refrigerators, which have as one of their design principles the recognition that they will be abused. Consequently, it is of prime importance that you read and follow instructions provided by the manufacturers of lab equipment before touching a new instrument. If questions arise direct them to the vendor or to the manufacturer. It is in their interest to have informed, satisfied customers; it is in your interest to know how to operate and care for your investment.

SUPPLIES

Following is a suggested list of initial supplies for a three-person lab (about 200,000 plants per year). Most of these supplies may be purchased from any scientific supply company; a few are available at supermarkets (see Appendix).

Item	*Size*	*Number*	*Total Cost (1986)*	*Use*
Alcohol, iso-propyl 78%	Pint	2	1.50	Clean explants and instruments
Aluminum foil, heavy	8 yd roll	1	1.50	Cover beakers, etc., when sterilizing
Beaker	100 ml	12	11.50	Clean explants
Beaker	250 ml	6	10.90	Clean explants, mix chemicals
Beaker (Fig. 9-b)	600 ml	4	13.80	Clean explants, mix chemicals, sterilize paper towels
Brush, flask	16″	1	3.00	Clean flasks
Brush, test tube	10½″	1	1.80	Clean test tubes, etc.
Canning jars (Fig. 11-e)	pint	1000	$450.	Grow cultures
Cotton nonabsorbent (Fig. 9-f)	1 lb.	1 pkg	8.05	Line top end of test tube closures
Culture tubes, "disposable" borosilicate (Fig. 9-e)	25 mm	12,800	1000.00	To grow cultures
Detergent, dishwasher, household	3 lbs	2	6.00	Dishwashing
Dishes, petri, glass	100×15 mm	12	30.96	Hold explant for micro-dissection
Dishes, plastic	2″×6″×9″	2	4.00	Hold bleach solutions in hood
Disinfectant (Lysol)	qt.	1	3.50	Clean benches, floors
Disinfectant, Sodium hypochlorite (Clorox, Purex)	1 gal	1	.85	Clean explants, instruments
Droppers, medicine	2 ml	12	1.75	Adjusting media pH
Erlenmeyer flask (Fig. 9-c)	1000 ml	6	26.04	Mix and store solutions
Erlenmeyer flask	2000 ml	1	11.09	Mix media
Erlenmeyer flask	4000 ml	1	29.23	Mix media
Filter, furnace	18″×25″×1″	6	6.00	Replace hood prefilters

Item	*Size*	*Number*	*Total Cost (1986)*	*Use*
Forceps, stainless steel (Fig. 9-g)	10″	2	26.60	Transferring cultures
Forceps, stainless steel, dissecting	4½″	2	9.66	Excising meristem
Gloves, household	As required	4 pr	8.00	Hand protection in hood
Grad. cylinder (Fig. 9-a)	10 ml	2	5.58	Measure liquids
Grad. cylinder	25 ml	2	6.60	Measure liquids
Grad. cylinder	100 ml	2	10.00	Measure liquids
Jar holder		1	3.00	Remove jars from canner
Knife blades, disposable (Fig. 9-g)	#10	100	28.00	Divide cultures and meristem
Knife holder, disposable blade, dissecting, stainless steel (Fig. 9-g)	#7	2	12.00	Hold blades
Knife, plastic handled, paring		3	1.00	Cut and divide cultures
Marking pens	med.	2	2.00	Labeling
Mop, household		1	10.00	Floor cleaning
Pipet	1 ml	2	5.00	Measure and dispense liquids
Pipet	5 ml	2	5.00	Measure and dispense liquids
Pipet	10 ml	2	6.00	Measure and dispense liquids
Pipet filler		1	16.51	Fill pipets safely
Pitcher	1000 ml	1	6.00	Dispense media
Polycarbonate	20 mil 2′ × 4′	10 sheets	80.00	Plastic for jar tops
Pot holders, kitchen	6″×6″	4	8.00	Hold hot items
Scoop	2 oz.	2	1.00	Scoop agar and sugar
Spatula	7″	2	6.49	Handle small amounts of chemicals
Sponge, household	medium	4	2.00	Cleaning
Test tube rack, 40, wire (Fig. 10-a)	For 25 mm tubes	4	46.00	Hold test tubes, instruments
Test tube rack, 10 (make from plywood) (Fig. 10-b)	12″	8		Hold test tubes when transferring and labeling
Thermometer	−20° to 110°C	2	23.47	General purpose
Timer, household	1 hr. in min.	1	8.00	Time media processing and explant treatments
Towels, commercial, single fold	9½″×10¼″	1 case	25.00	To support cultures on hood work surface
Towels, household, paper	Roll	6	3.00	General purpose
Trays, Todd™ planter, or Speedling™ (Fig. 10-c)	For 128 test tubes, 13⅝″× 26⅝″×2½″	42	144.92	To hold test tubes in growing room
Tube closures Kim-Kaps™ or Ka-Puts™ (Fig. 9-f)	For 25-mm tubes	12,800	600.00	To cap tubes

Item	*Size*	*Number*	*Total Cost (1986)*	*Use*
Wash bottle (Fig. 9-d)	250 ml	4	8.00	Rinse probes, beakers, etc., apply small amounts of liquids
Wash bottle	500 ml	4	10.40	Rinse beakers, etc., water plantlets in Stage IV
Wastebasket	Large	1	6.00	Waste in media room
Wastebasket	Medium	1	4.00	Waste in transfer room
Weighing papers	3″×3″	1 pkg	5.15	Hold chemicals on balance

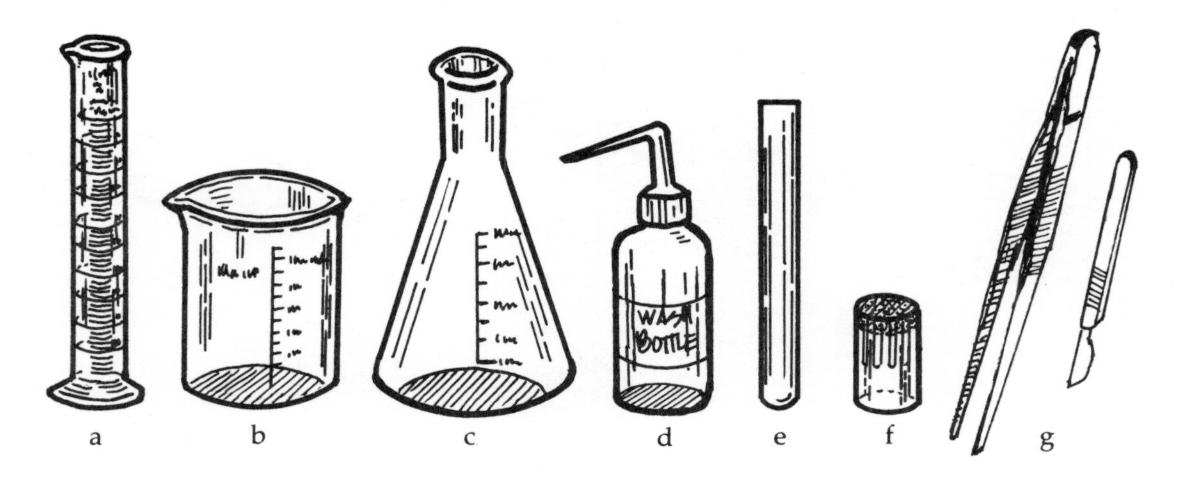

Figure 9. (a) Graduated cylinder. (b) Beaker. (c) Erlenmeyer flask. (d) Wash bottle. (e) Test tube. (f) Closure (cap) for test tube, showing cotton filter. (g) Forceps and knife (scalpel).

NOTES ON SUPPLIES

Pint Mason jars are readily available, convenient, autoclavable containers in which to tissue culture plants. Used upright, they are not too deep to position plantlets and are easily handled. The metal tops do not allow sufficient light to the cultures, so it is necessary to substitute some type of transparent lid. Rounds of plate glass are expensive and cumbersome. Polycarbonate is an autoclavable plastic available in 20 mil sheets and suitable for easily cutting rounds to fit the tops of pint jars. When processing media, use the metal rings (which come with the jars) to hold the rounds in place. After placing the transfers into the jar, replace the polycarbonate top (which was set on a sterile paper towel) and place a square of sandwich wrap plastic over it ot hold it on. The right size square is obtained by cutting a whole roll of plastic wrap in two, making two short rolls; tear off a square at a time. Press the plastic wrap square over the top and to the jar and fasten with a rubber band close to the top (Figure 11-e), or use the metal band over the wrap. In very clean areas the wrap is no advantage; simply replace the polycarbonate top and the metal ring.

Test tube closures do not offer maximum protection against contaminants. Line them on the top end with ¼″ thick piece of non-absorbent cotton as illustrated (Fig. 9-f).

An instrument holder is useful in the hood and required when using a Bacti-Cinerator (Fig. 8-2). A holder can be made quite easily. Take a piece of ¾″ hardware cloth, 6″ × 22½″. Form so that the back is 10½″ high, the front 9¾″ high, and the base between is 2¼″ wide. Place in a mold 2″ deep and pour in wet Plaster of Paris. When dry remove from the mold. A quicker, more expensive holder is a metal test tube holder standing on end. Do not buy a plastic-coated one. A zinc-plated, steel wire rack 10⅜″ × 4⅜″ × 3¼″ costs $12.15. A stainless steel rack costs more.

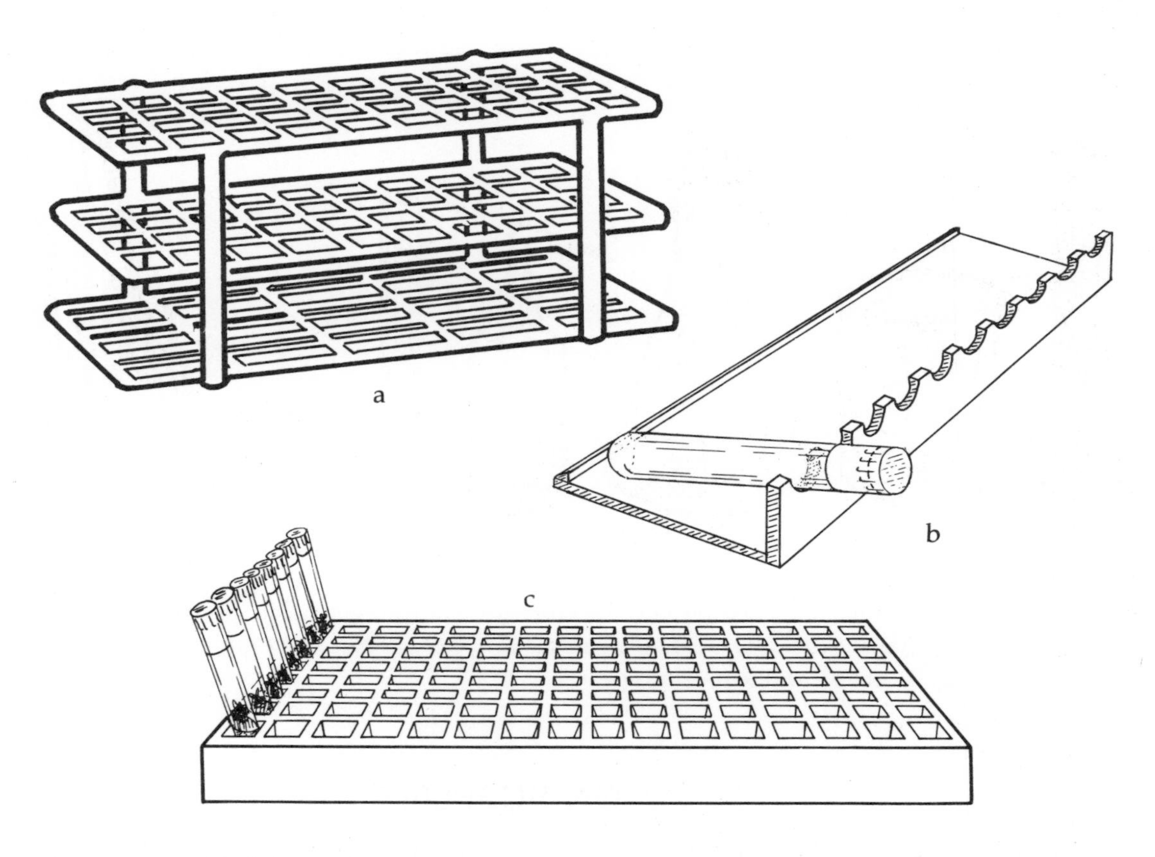

Figure 10. (a) Wire rack test tube holder for 40 tubes. (b) Plywood test tube holder for 10 tubes. (c) Speedling or Todd planter tray used as test tube holder for 128 tubes.

CLEANLINESS

Having described the requirements for building and outfitting a lab, with cleanliness as a major planning factor, it is time to dwell on some procedures for maintaining cleanliness and put them in perspective before launching into tissue culture methodology.

It is not practical to strive for the sterility of an operating room, but cleanliness is extremely important in a tissue culture lab. The cleaner the lab the less opportunity there is for contamination. Contaminants are bacteria, mold spores, yeasts, etc., which enter cultures by way of the air which is allowed into containers, or by means of unsterile implements or containers. Contaminants grow well on tissue culture media, consequently overgrowing and killing the cultures, or, at least, impairing their growth. Prevention is the only control against contaminants because there are no satisfactory cures.

There is a foreign custom of removing shoes before entering a home, a very good practice to establish with a laboratory. Workers can supply their own footwear to be worn only in the lab. Visitors are very understanding when asked to remove their shoes. An alternative is to offer visitors plastic sacks to wear over their shoes, or disposable shoe covers which can be purchased from scientific or safety supply companies. Even with such precautions, a vinyl floor rapidly accumulates a host of contaminants. A wet sponge mop using a Lysol solution should be used to clean the floor daily, never a broom, vacuum, or dry mop because these stir up dust. Walls and counters should be washed at least weekly with a disinfecting solution. The hood prefilter (Figure 8-1-e) should be changed every four months.

Personal cleanliness is very important; long hair should be secured so it will not contaminate, or catch fire, in the hood. Clean clothing and fresh lab coats contribute to cleanliness and good morale.

References: #56, #82, #139, #151, #160, #162.

Chapter III: Preparing Media— The "Kitchen" in Action

This chapter deals with the preparation of chemicals that support the life of plant tissue cultures, stimulate their differentiation, and guide their growth. Actually, the situation is little different from preparing a menu; the ingredients and language are different, and the equipment is perhaps unfamiliar, nevertheless, recipes are followed in a precise, orderly, meaningful fashion.

For those of us who may have been bored with high school or college chemistry the elements take on a new dimension, relating and translating into plant response which can be nothing short of dramatic. Skim or read the chapter for what it can mean to you. If a cupboard full of unfamiliar ingredients is too overwhelming then buy "premixed" media—at least to start with. If the chemistry review is meaningful to you, then I urge you to take up the challenge because the whole world of micropropagation is open to you. As a grower you will not be entirely in new country because you have used fertilizers and rooting hormones. Start with the ball wherever you are and see how far you can run with it.

CHEMICALS

It is difficult, but not impossible, for someone who has not had at least a beginning course in chemistry, to tackle the chemical language, symbols, and working principles of mixing media. Initially the number of chemicals used to produce media may seem overwhelming. However, if the beginning tissue culture propagator starts with a plant for which media formulas have been well researched and applied, the task becomes relatively simple. A cookbook approach fortified with determination will help the novice gain confidence.

An alternative to mixing numerous chemicals is to purchase ready mixed, powdered media which are available from several companies. In spite of premixed media, however, labs usually choose to do their own mixing of formulas not only because it is cheaper but also they can modify standard formulas in the light of accumulated experience.

In any case it is useful for the technician to know something about chemicals and their role in plant growth and development. The information presented here should be both useful and helpful to the beginner to appreciate the art and science of media formulation.

Many of the following terms and concepts that are used by chemists are household words; however, a brief review of definitions is necessary to strengthen the foundation we are laying for formula mixing.

The metric system

The system of measurements used in science, in tissue culture, and in the lab is the metric system. It is not difficult to become accustomed to this system and calculations are much easier than in the English system because it is all in multiples of 10. A few relationships and terms committed to memory will make metric calculations and measuring work much easier for the person trained in the English system. To mention a few: there are 1000 mg (milligrams) in a gram, 1000 ml (milliliters) in a liter, 10 mm (millimeters) in a cm (centimeter), and 100 μ (microns) in .1 mm–about as small as the unaided eye can see. (See Appendix.)

Elements, atoms, molecules, and compounds

An *element* is a substance that cannot be separated into different substances by any usual chemical means. An atom is the smallest particle of an element that retains the characteristics of that element. In a free state atoms exist as molecules, sometimes as single atoms (N, P, or K, for example), and sometimes a molecule of an element consists of two atoms (H_2, or O_2, for example). The number of atoms is always written as a subscript, below and to the right of the letter symbol.

A *compound* is two, or more, different elements chemically combined in fixed proportions, for example, water, H_2O; hydrochloric acid, HCl; or sucrose, $C_{12}H_{22}O_{11}$. A *molecule* of a compound is the smallest part of that compound that can exist in a free state and still retain the properties of that compound.

An *ion* is an electrically charged atom or group of atoms. For example, table salt, sodium chloride (NaCl) ionizes in water to form positively charged sodium ions(Na«) and negatively charged chloride ions (Cl^-). When water containing calcium carbonate (Ca«CO_3) is deionized the positive calcium ions (Ca^{++}) are removed by a negatively charged resin bed and the negative carbonate ions (CO_3^{--}) by a positively charged resin bed.

Atomic and molecular weights

The occasional need to use atomic and molecular weights warrants an introduction to the topic. An atom is the smallest particle of an element that can combine with other elements, an element being a substance that cannot be separated into different substances by any common chemical means. The atoms of different elements have different weights. When atoms of different elements are chemically combined in fixed proportions they form compounds. The smallest part of a compound that can exist in a free state and retain the characteristics of that compound is a molecule. The sum of the atomic weights in a molecule is the molecular weight (MW) of the compound.

Sometimes formulas are stated in molar quantities rather than in grams per liter. A one molar solution is one gram molecular weight (the molecular weight of a substance expressed in grams, also called a mole) in a liter of water. For example, a one molar solution of sodium hydroxide (NaOH) is often used to raise the pH of media. Look at the table of atomic weights in the Appendix. The atomic weight of sodium (Na) is 23, oxygen (O) is 16, and hydrogen (H) is 1; therefore the molecular weight of NaOH is 40. Forty grams of NaOH in a liter of water equals a one molar (1M) solution of NaOH. Molecular weights are often found in the commercial catalogs of chemical companies. (Aldrich) These are especially helpful where organic chemicals are concerned.

Another instance where an understanding of atomic weights is useful is with respect to the water of hydration. Water of hydration is the variable amount of water that is chemically attached to some compounds. Manganese sulfate ($MnSO_4$) is manufactured with either 4 molecules of water ($MnSO_4 \cdot 4H_2O$, MW = 223) or as the monohydrate ($MnSO_4 \cdot H_2O$, MW = 169). If 2.23 grams of $MnSO_4 \cdot 4H_2O$ is called for, 1.69 grams of $MnSO_4 \cdot H_2O$ ($169 \div 223 \times 2.23 = 1.69$) would provide an equivalent amount of $MnSO_4$ when using the monohydrate.

Media formulas, published in technical journals, are sometimes specified in millimoles (1/1000 gram molecular weight) per liter instead of grams per liter. In this case it is necessary to find the molecular weight of the substance and multiply the moles required times the molecular weight (MW). For example if 3 mM (millimoles) of $CaCl_2$ are required per liter, the MW of $CaCl_2$ times the millimoles required equals the milligrams of $CaCl_2$ required per liter of media, 110.98×3 mM $= 333$ mg. The advantage of expressing media ingredients in moles is that molecules and ions interreact as entities; thus, comparing moles instead of weight provides a more valid basis of comparison of one formula with another.

Weights and measures

When measuring chemicals it is useful to be aware of the concept of *significant figures.* Simply stated, when weighing chemicals, the larger the amount to be weighed the less important is precise accuracy. A milligram can be lost and is insignificant in 100 grams, but when only 10 milligrams is required a milligram is a significant factor amounting to 10%. This degree of importance is referred to as the concept of significant figures; it applies to both weights and volumes. Three significant figures in the final measure are adequate for production work; thus 184 grams (three figures) is sufficiently accurate when 184.4 grams are called for; or 15.6 milligrams is entirely satisfactory rather than 15.57. This point is made because often in the literature more than three figures are quoted. People doing research must be extremely precise, while people in production need not be quite so accurate. The concept is also useful to employ when adjusting formulas and the calculator spins out seven decimals during the calculations.

Some laboratory glassware, particularly beakers and Erlenmeyer flasks, carry the caution that the calibrations on the side are accurate only to within plus or minus 5%. Accuracy within 5% is usually sufficient for production purposes.

Principles of acids and bases—solutions for pH adjustment

pH is a symbol for the degree of acidity or alkalinity of a solution as indicated by the hydrogen ion concentration. Different plant species require different pH for optimum growth.

A solution is neutral at pH 7, alkaline above pH 7, and acid below pH 7. The numbers are exponential; thus a solution at pH 6 has ten times the hydrogen ion (H^+) content as a solution registering pH 7, a solution at pH 5 has 10 times as many H^+ ions as a solution at pH 6, etc. In the opposite direction, a solution at pH 8 has 10 times the hydroxyl (OH^-) ions as a solution at pH 7, etc. The amount of hydroxide or acid required to change pH a degree varies considerably, depending upon the nature of the solution and at what point on the scale the change is made.

The usual pH range for tissue culture media is between pH 4.5 and pH 5.7. (Table 1)

Table 1. Range of pH
(Plant tissue culture media: 4.5–5.7)

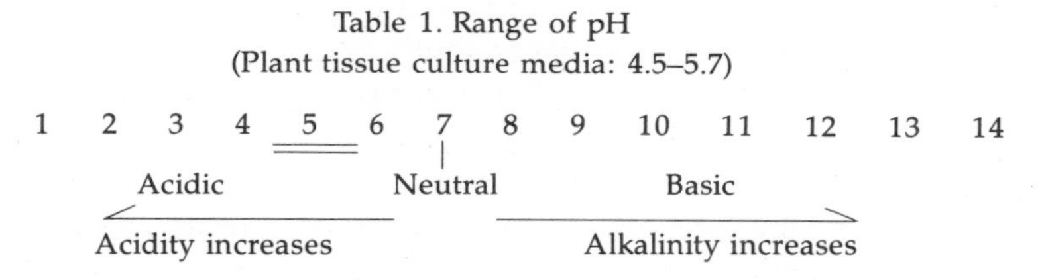

In general, plant response is not too sensitive to pH over a small range. (Occasionally, however, the pH of a medium after it has grown plant material for a period of time will show a significant change that is detrimental to the culture.)

Sodium hydroxide (NaOH) solutions are used for raising the pH of media (making them more alkaline) when they are too acidic. The NaOH solutions contain sodium (Na^+) ions and hydroxyl (OH^-) ions. The OH^- ions combine with the excess H^+ ions in the medium to form water (H_2O), thus neutralizing them and causing the medium to be more basic (alkaline). NaOH is usually purchased in the form of pellets. Do not handle this chemical without using tweezers, spoon, or spatula as it is corrosive. To 100 ml of water add 4 g of NaOH for a one molar (1M) solution. Use great care when making this solution because when NaOH and water mix it is extremely reactive and may spatter. The caustic spatter will injure eyes, skin, or anything it touches. One molar (1 normal) NaOH may be purchased.

Hydrochloric acid (HCl) solutions are used for lowering the pH of media (making them more acid) when they are too alkaline. The HCl solution contains hydrogen (H^+) ions and chloride (Cl^-) ions. The H^+ ions combine with the excess OH^- ions in the medium to form water, thus neutralizing them and causing the medium to be more acidic.

To lower the pH of media use a 1 molar solution of hydrochloric acid (1M HCl). The molecular weight of HCl is 36.5 and it often is purchased as a 38% liquid; therefore, 9.6 grams of HCl is required for 100 ml of 1 molar HCl ($9.6 \times .38 = 3.65$ g HCl). To make 100 ml of 1 molar HCl tare (weigh) a beaker containing 75 ml of water and slowly add HCl with a pipet until the balance indicates you have added 9.6 grams of HCl (weight of beaker and water plus 9.6 g HCl). Never add water to strong acid because it will spatter and injure anybody or anything it touches. After adding the HCl, remove the beaker from the balance and carefully add water to make 100 ml of solution. Of course it is easier to buy 1M (1 normal) HCl already prepared.

How these solutions are used will be described in the section on the media making process.

There will be more terms and concepts defined as we proceed but the foregoing will be useful in the following discussion of chemicals.

INORGANIC CHEMICALS

Growers will immediately recognize the essential elements all plants require and not be surprised to find them in tissue culture media. Seven of these are major constituents of common fertilizers: nitrogen (N), phosphorus (P), potassium (K), sulfur (S), calcium (Ca), magnesium (Mg), and iron (Fe). (The three remaining major elements—carbon (C), hydrogen (H), and oxygen (O)—will be discussed later under organic chemicals where they play a major role.)

MAJOR ELEMENTS

- *Nitrogen* (N) influences the rate of plant growth. It is an essential element in the molecular make-up of chlorophyll, alkaloids, nucleic acids, some plant hormones, and amino acids. Lack of nitrogen is characterized by yellowing leaves and a stunting of growth. Excess nitrogen promotes vigorous growth but suppresses fruit development. Sources of nitrogen for tissue culture media are ammonium (NH_4^+) and nitrate (NO_3^-) compounds.

  The concentration of salts can have a profound effect upon culture performance, particularly the level of ammonium ions. Reduction of MS formula ammonium nitrate (NH_4NO_3) and potassium nitrate (KNO_3) to ¼ strength was found essential in a report on bud induction of Douglas fir. Studies of other genera (*Rhododendron, Lycopersicon, Arabidopsis,* and *Torenia,* to mention a few) have noted similar effects.
- *Phosphorus* (P) is abundant in meristematic and other fast growing tissues, but its exact role is not known. Its chief purpose appears to be that of an enzyme activator (enzymes are compounds which promote reactions without being part of them). Too little phosphorus causes plants to be abnormal and sickly. Potassium phosphate (KH_2PO_4) and sodium phosphate ($NaH_2PO_4 \cdot H_2O$) are routinely included in tissue culture media.
- *Potassium* (K) is necessary for normal cell division, for synthesis of carbohydrates and proteins, for manufacture of chlorophyll, and for nitrate reduction. Insufficient potassium results in weak and abnormal plants. Potassium nitrate (KNO_3) and potassium phosphate (KH_2PO_4) are common sources of potassium in tissue culture media; potassium chloride (KCl) is used occasionally.
- *Sulfur* (S) is present in some protein molecules. It promotes root development and deep green foliage. It is supplied in tissue culture media in sulfate (SO_4^{--}) form.
- *Calcium* (Ca), as calcium pectate, is an integral part of the walls of plant cells where it plays a role in permeability. It facilitates the movement of carbohydrates and amino acids throughout the plant and promotes root development. As calcium oxalate it ties up oxalic acid, a by-product of protein metabolism. Calcium is usually included in tissue culture media as calcium chloride ($CaCl_2 \cdot 2H_2O$) or as calcium nitrate ($Ca(NO_3)_2 \cdot 4H_2O$).
- *Magnesium* (Mg) is the central element in chlorophyll molecules. It is also important as an enzyme activator. Magnesium deficiency causes pale, sickly foliage. Most tissue culture formulas call for magnesium sulfate ($MgSO_4 \cdot 7H_2O$, also known as Epsom salts).

- *Iron* (Fe) is involved in chlorophyll synthesis. It also participates in energy conversion in photosynthesis and respiration as it is reduced from the ferric to the ferrous state. Plants deficient in iron are chlorotic (pale, yellowed). In tissue culture media, ferrous sulfate ($FeSO_4 \cdot 7H_2O$) is mixed with the sodium salt of ethylene-diamine tetraacetic acid (EDTA) which renders the iron more readily available to the plants.

MINOR ELEMENTS

In addition to the major elements required by plants there are a number of other elements essential to good growth but which are needed only in extremely small quantities. They are called minor elements, trace elements, or micronutrients. Their functions are not well understood. In fact, only recently have they been identified as necessary and some of the associated deficiency symptoms recognized. When the purity of the water and chemicals used in tissue culture media reached the refinements of today's capabilities it was discovered that trace elements had previously been supplied as undetected impurities in presumably pure water and chemicals. Trace elements are present in soil, water, and even dust particles, in adequate amounts for plant growth. Some "chemically pure" compounds used in making media may contain traces of these elements; if they do, they are listed as impurities on the labels. Several trace elements are toxic to plants in excess amounts.

- *Boron* (B) is an important trace element presumed to play a role in sugar movement. Lack of boron produces interesting deficiency symptoms, usually deterioration of internal tissues as in "heart rot" of sugar beets, cracked stem of celery, or "monkey face" in olives. Excessive amounts of boron cause plant injury or death, thus some herbicides are borates. Boron is added to tissue culture media in small amounts as boric acid (H_3BO_3).
- *Molybdenum* (Mo) is believed to participate in the conversion of nitrogen to ammonia; it also aids nitrogen fixation (the conversion of atmospheric nitrogen to nitrates by nitrogen fixing bacteria). It is required for normal growth and is added to tissue culture media as sodium molybdate ($Na_2MoO_4 \cdot 2H_2O$). Quantities exceeding 10 parts per million can be injurious to plants.
- *Manganese* (Mn) deficiency is characterized by various chlorotic symptoms, often yellow mottling of leaves. It is an essential element in the chloroplast membrane. Manganese sulfate ($MnSO_4 \cdot H_2O$) supplies the necessary manganese in tissue culture media.
- *Cobalt* (Co) is an element in the complex vitamin B_{12} molecule ($C_{63}H_{90}N_{14}O_{14}PCo$) which is essential to nitrogen fixation. Cobalt chloride ($CoCl_2 \cdot 6H_2O$) is added to most media at .025 mg per liter.
- *Zinc* (Zn) is a vital element in several enzymes; it is involved in chlorophyll formation as well as in the production of the auxin, indoleacetic acid (IAA). A trace of zinc as zinc sulfate ($ZnSO_4 \cdot 7H_2O$) is included in tissue culture media; large quantities are toxic to plants as is true of most trace elements.
- *Copper* (Cu) deficiency causes stunted growth, malformations, twisted and blotched leaves, or die-back of young twigs. Copper is believed to be necessary in energy conversion as it alternates between the cuprous and the cupric state. Only .025 milligrams of cupric sulfate ($CuSO_4 \cdot 5H_2O$)

per liter of media is required to supply the necessary copper in tissue culture of most plants.

- *Chlorine* (Cl) is essential to help stimulate photosynthesis; how it works is not understood. Deficiency symptoms are wilted leaves which become yellowed or bronze and die. Plants require chlorine in minute quantities, but it is included in some tissue culture media in large amounts as calcium chloride ($CaCl_2 \cdot 2H_2O$). Excess amounts of this micro-nutrient appear acceptable to most plants, but should be suspect in some cases.
- *Iodine* (I) is added to media as potassium iodide (KI). It is not usually considered an essential element even though it is a component of some amino acids. Some labs omit iodine from rhododendron medium.

ORGANIC CHEMICALS

Organic chemicals are compounds containing carbon, such as carbohydrates, hormones, proteins, enzymes, etc. They are not usually provided to plants because plants normally manufacture their own. However, plants in culture are either too small or too incomplete to synthesize all of the organic chemicals they need so organic substances must be added to the tissue culture media to augment the plantlets' autotrophic (self-generated) supply.

Carbohydrates are organic chemicals such as sugars, starches, and celluloses. In varying amounts and configurations carbon (C), hydrogen (H), and oxygen (O) are the primary elements that make up the molecules of these compounds. These elements are generously supplied by air and water primarily as carbon dioxide (CO_2) and wat (H_2O).

- *Sucrose* ($C_{12}H_{22}O_{11}$) is a sugar found abundantly in plant tissues. It is composed of two simpler, chemically combined sugars, fructose and glucose. As a carbohydrate it is an indirect product of photosynthesis, the process which, with the help of chlorophyll and light, converts carbon dioxide and water into carbohydrates and releases oxygen. But plants growing in tissue culture media cannot make all the sugars they require, so a high concentration of sucrose, 30 grams per liter, is specified in most media formulas. Cane or beet sugar, both virtually 100% pure sucrose, as purchased from the grocery store, are good sucrose for tissue culture media.

Vitamin B complex contains essential compounds for plant metabolism and growth. The growth substances in yeast extract, historically used in media in the past, are now identified as thiamine, nicotinic acid and pyridoxine, all members of the vitamin B complex. The following vitamins of the B complex have been found to be desirable inclusions in some tissue culture media.

- *Inositol* ($C_6H_{12}O_6$) is one of the B complex required in most media. Described as a "sugar alcohol," in its phosphate form it is part of various membranes, particularly those of certain organelles such as chloroplasts. Inositol (myo-inositol) is added to tissue culture media in relatively large amounts, 100 milligrams per liter.
- *Thiamine* (Vitamin B_1, $C_{12}H_{17}ON_4SCl$) functions in a coenzyme to assist the organic acid cycle of respiration (Krebs cycle). Only .4 milligrams of thiamine hydrochloride per liter is specified for many tissue culture media.

- *Nicotinic acid* (niacin, $C_6H_5O_2N$) is a constituent of coenzymes active in light energy reactions. Various media formulas specify nicotinic acid ranging from .1 to 10 milligrams per liter.
- *Pyridoxine* (Vitamin B_6, $C_8H_{11}O_3N$) also serves as a coenzyme in metabolic pathways (chemical reactions of metabolism). As the hydrochloride form it is sometimes included in culture media.
- *Pantothenic acid* ($C_9H_{17}O_5N$), another B vitamin, is active as a coenzyme in fat metabolism. It is occasionally added to media as the calcium salt.
- *Folic acid* ($C_{19}H_{19}N_7O_6$) is found in green leaves and other plant tissue. It functions as a vitamin B demonstrating coenzyme activity. It is occasionally used in media.
- *Choline* ($C_5H_{15}O_2N$) is an alkaloid within the vitamin B complex. Choline chloride is occasionally specified in media formulas.
- *Riboflavin* ($C_{17}H_{20}O_6N_4$) is known as Vitamin B_2 and is occasionally used in media.

Vitamin H (biotin, $C_{10}H_{16}O_3N_2S$) is active as a coenzyme in fat metabolism. It is occasionally used in media.

Growth regulators, or hormones, are placed in two categories: auxins and cytokinins. Auxins promote cell enlargement and root initiation. Cytokinins promote cell division and shoot initiation. This is a simplistic summary, in view of the diversity of compounds which are included in these two categories of growth regulators. There is a wide range of interaction between these groups; they also interact with other chemicals and are affected by environmental factors, such as light and temperature. It has been suggested an auxin might even react as a cytokinin and a cytokinin as an auxin in particular conditions. In any case, it is of the greatest importance that tissue culture media contain the correct ratio and the right kind of growth regulators for each species of plants being cultured.

- *Auxins.* F. W. Went (1926) found that a substance extracted from oat coleoptile (young grass shoot sheath) tips caused bending in other coleoptiles from which the tips had been removed. Kagl (1934) and Thimann (1935) identified an auxin, indoleacetic acid (IAA), as the same substance Went had discovered.

  With the subsequent synthetic production of indolebutyric acid (IBA) and naphthaleneacetic acid (NAA), further experiments revealed that these substances are even more effective than IAA for cell enlargement and root initiation. IBA and NAA have found widespread use in the induction of roots on cuttings. Auxins are commonly used in tissue culture media, combined with cytokinins during the multiplication stage and, without cytokinins, during the rooting stage.
- *Cytokinins,* formerly called kinins, are growth regulators that promote cell division, help control seed germination, affect leaf abcission, influence auxin transport, allow gibberellins to work by overcoming inhibitors, and delay scenescence—for example, they postpone the breakdown of chlorophyll, protein, and nucleic acids in detached leaves.

  F. C. Steward launched a search for cytokinins prompted by the dramatic growth effect of coconut milk on carrot root cultures, often an increase of 80 fold in three weeks. His tests led to the discovery of growth factors in various endosperm including the cytokinin, *zeatin,* in corn seed endosperm. *Kinetin* was extracted from degraded herring sperm DNA (the primary material in chromosomes, *deoxyribonucleic acid*) by C. Miller in Skoog's lab in Wisconsin in 1955.

Iso-pentenyl adenine (2iP, also called 6-(y,y-dimethylallyl) amino purine) has been found in RNA (*ribonucleic acid,* a carrier of genetic information) and in a pathogenic bacterium *(Corynebacterium fascians)*; it causes rapid cell division and consequent irregular growth in some higher plants. Assumed to be widespread in plants, but difficult to find, kinetin, 2iP, and *benzyladenine* (BA, or benzylaminopurine, BAP) are manufactured synthetically.

Cytokinins are required in tissue culture media for cell division and shoot multiplication; they are usually omitted in media for the rooting stage. If cultures are too long and skinny, increased cytokinin will increase multiplication and decrease length.

- *Adenine* ($C_5H_5N_5$) is important to cells as part of the nuclear substances. In practical experience it has a weak cytokinin effect. As adenine sulfate it is included in some tissue culture media formulas to help promote shoot growth.
- *Gibberellins* are a group of naturally occurring substances that influence cell enlargement. Abnormally rapid growth of rice seedlings due to a fungal secretion was noted by Kurasawa in 1926. The substance was gibberellic acid (GA_3) which later was isolated in crystalline state from both fungi and higher plants. Thirty-four gibberellins have been chemically identified. Some of them appear in embryos, where they initiate production of alpha amylase, which converts starch to sugar and stimulate other enzymes. GA_3 is often used as a supplemental growth regulator to auxins and cytokinins in culture media.
- *2,4-dichlorophenoxyacetic acid* (2,4-D) is well known as a weed killer. It has been widely used in plant tissue culture media, especially to induce callus.
- *Growth retardants* appear to have a cytokinin effect in tissue culture. *Ancymidol* (a-cycloprophyl-a(p-methoxyphenyl)-15-pyrimidine methanol), available as A-Rest, has been found to shorten shoots and induce buds in some cultures.
- *Phloroglucinol* has been suggested as an aid in prevention of vitrification.

Amino acids are building blocks of proteins and nucleic acids. They are not routinely added to media but their occasional use implies value in specific cases.

- *L-Cysteine* ($HSCH_2CH(NH_2)COOH$) is an amino acid occasionally specified in media formulas and added as the hydrochloride.
- *Glycine* ($C_2H_5O_2$) is a simple amino acid occasionally called for in media formulas.

Antibiotics

- *Agrimine* (Reichhold Chemicals, Inc.) While no longer available from Reichhold, it has been found useful against certain yeasts. Described by Reichhold as a 10-0-0 fertilizer with nitrogen derived from urea-formaldehyde reaction products, it is available elsewhere as methylolurea solution or UF-85 concentrate. Autoclavable.
- *Cefotaxime.* Used to eliminate certain woody plant bacteria. Use at 25 mg/L in combination with tetracycline, rifampicin, and polymyxin-B. Cold filter sterilize.

- Polymyxin-B. Used to eliminate certain woody plant bacteria. Use at 25 mg/L in combination with cefotaxime, tetracycline, and rifampicin. Cold filter sterilize.
- *Rifampicin.* Used to eliminate certain woody plant bacteria. Use at 6 mg/L in combination with polymixin-B, cefotaxime, and tetracycline. Dissolve in DMSO (dimethylsulfoxide). Cold filter sterilize.
- *Tetracylcine.* Used to eliminate certain woody plant bacteria. Use at 6 mg/L in combination with cefotaxime, rifampicin, and polymyxin-B. Cold filter sterilize.
- *Streptomycin.* Use at 20 mg/L. Often helpful even when autoclaved.
- *Gentamycin.* Described as an autoclavable antibacterial. Occasionally effective.
- *Ribivirin* (Virazole) A broad spectrum antiviral agent. Used effectively against some potato viruses.

Undefined constituents of tissue culture media

- *Coconut milk* is an undefined medium used with success in early plant tissue culture and still used today. It was first reported by van Overbeek (1941) in tissue culture of *Datura* embryos. Caplin and Steward (1948) grew carrot phloem explants using coconut milk and casein hydrolysate as basal salt supplements. This medium produced far more callus than did a supplement including zeatin, inositol, casein hydrolysate, and IAA. Coconut milk has been used widely in orchid culture; Withner (1961) first cultured *Phalaenopsis* on Vacin and Went medium which included 15% coconut milk.
- *Casein hydrolysate* is an undefined protein mixture occasionally used in media. Casein, from milk, is hydrolyzed (treated with water) to form this weak acid.
- *Yeast extract* is a natural source of vitamins. It is purchased as a powder and occasionally used in media. (See *Prunus* media, pages 122–23.)
- *Agar* is a mixture of polysaccharides derived from extracts of several species of red algae. (A polysaccharide is a carbohydrate that has a large number of simple sugars, such as glucose, linked together in its chemical structure.) As a gelling agent, the agar in tissue culture media is strong enough to support the culture yet liquid enough to allow the nutrients to diffuse through the medium to the plantlets.

  Although agar is supposedly an inert material in tissue culture media it frequently contains traces of other elements. The concentration of impurities in agar varies with the source of the algae and the method of manufacture. One analysis revealed a range of chloride content, varying from .31% in Spanish sources to only a trace in Japanese and Portuguese sources. Other impurities discovered in some agars include sulfate, calcium, magnesium, and iron ions. "Purified" agars sold by tissue culture suppliers are usually sufficiently pure for plant tissue culture.

  Gel strength was found to vary by 35% between the lowest and the highest gels in a group of agar samples from different sources. Six grams of agar per liter of media is usually satisfactory but gel strength will vary both with the medium formula being used and the source and grade of agar. The gel formed should be firm enough that tubes will hold a good slant without being sloppy, yet soft enough that plant material can be pressed gently and easily into good contact with the agar. Media with low salts or hormones will tend to be harder than with full salts and hormones. Media with a low pH (pH 4.5) will tend to be softer than media with higher pH (pH 5.7). To some extent ideal gel strength varies with the plant cultivar. There is little research on the subject but there are indications that gel strength sometimes affects culture performance.

Several agar substitutes have been developed. One such gelling agent, Gelrite, is derived from bacteria (a *Pseudomonas sp.*) and is available commercially (Kelco). Similar to agar, it is a highly refined polysaccharide. In our experience, media made with it is exceedingly clear and no ill effects in the cultures have been observed; however, we prefer to mix it with agar as a 3 Gelrite/1 agar ratio. The price is twice that of "purified" agar, but only one third as much Gelrite is necessary to provide the same gel strength.

- *Activated charcoal* is frequently added to rooting media. Its function is not clear but it appears to absorb root inhibiting agents, possibly cytokinins among others. It is usually added at 0.6 grams per liter.

MEDIA FORMULAS

Through trial and error, research scientists have discovered various chemical combinations (formulas) that will make certain plants grow and multiply in culture. For the most part they have published their results in technical journals that are not readily available to most growers. For this reason, tissue culture media formulas (recipes) for 47 cultivars are provided in Section II of this book.

Standard media formulas, as used here, are formulas that have been determined by research scientists to provide optimum nutrients and growth regulators for specific cultivars. It should be noted that a cultivar will often grow and multiply in a medium designed for another species but may not perform at its optimum.

Probably the best known standard formulas are those developed by Murashige and his associates, MS in particular (pages 17, 109). They are primarily designed for herbaceous foliage plants such as *Nicotiana,* gerbera, ferns, begonia, lilies, etc. Murashige described three stages of culture and prescribed specific media for each stage of each variety. Stage I is the period of explant establishment; Stage II is the multiplication stage; and Stage III is the culture in which rooting is induced. A Stage IV is referred to by some authors. It is simply the stage when plantlets are transferred out of *in vitro* culture to artificial soil medium (potting mix). If plantlets are rooted directly into soil, as if they were miniature cuttings, Stage III is omitted.

A formula developed by Brent McGown and Greg Lloyd at the University of Wisconsin is called "Woody Plant Medium" (WPM). This formula was designed to optimize tissue culture of certain woody plants as the name implies. With minor variations WPM is used to culture a wide variety of woody plants including *Kalmia,* birch, rose, *Rhododendron,* and oak. The main difference between WPM and MS is the relatively small amount of sodium and chloride ions in WPM. Further, WPM prescribes only 75% of the ammonium and nitrate ions and 60% of the potassium recommended in MS.

There are two main routes by which multiplication is brought about. The specific nutrients used determine the evolving pathway. The simpler route promotes lateral and adventitious shoots directly, beginning with the explant. The other route is to first induce the explant to produce callus (unorganized) tissue. The callus cells are multiplied then conditioned to produce embryoids, shoots, and roots. Callus has research advantages but is cumbersome for commercial micropropagation. Historically, most early research followed the callus route. Because of new discoveries more and more plants can be cultured without a callus phase. The direct shoot route will be followed almost exclusively in the pages to follow.

Premixes

For the grower who does not wish to mix numerous chemicals there are several tissue culture supply companies which sell premixed, powdered media both in various formulations and degrees of completion. Examples of available premixed media are:

Murashige and Skoog (MS) salt base medium. This is a versatile, inorganic compilation to which one may add any organic chemicals desired. It is also good, without additions, for starting many explants.

MS shoot multiplication medium. This is a full formula which is good to try on herbaceous material where no formula is available that is specific to the plant in question.

Murashige formulations, complete and specific for ferns, gerberas, cordyline, dracaena, African violet, etc.

White's medium.

Vacin and Went medium for *Cymbidium* orchid multiplication.

Nitsch's medium.

These and other premixes are available with or without sugar, agar, or hormones (cytokinins and auxins). The more complete formulas even have the pH preadjusted. If a grower desires a premixed formula that is not already commercially available, some supply companies will custom make powdered premixes to order. (See Appendix, p. 137.)

The cost effectiveness of buying premixed media must be decided by the individual lab in terms of the various cost factors involved. Premixes do, of course, have to be added to water, heated to dissolve agar, supplemented, and pH adjusted where necessary, and dispensed. Buying total formulas as powdered premixes is convenient but limits the opportunity to vary the ingredients to manipulate and optimize culture growth for specific species or cultivars. On the other hand, buying premixed salts allows the leeway to add the remaining ingredients for specific needs and also eliminates the chore of making and storing stock solutions of salts.

Quick and Easy

For the person who thinks tissue culture is all too complicated and "scientific," Bridgen and Brand (University of Connecticut) propose the ultimate do-it-yourself strategy using the grocery store, pharmacy, and health food store as sources of medium ingredients. They suggest the following off-the-shelf recipe:

Table sugar	1/8 cup
Tap water	1 cup
Stock: 1/4 tsp. all purpose 10-10-10 fertilizer in 1 gal. water	1/2 cup stock
Inositol tablet (250 mg)	1/2 tablet
Vitamin tablet with thiamine	1/4 tablet
Agar flakes	2 tablespoons

Boil until the agar has melted, stirring continuously. Dispense into pint canning jars or babyfood jars so that medium is ½" to 1" deep. Cover and process in a pressure cooker, according to directions of the cooker manufacturer, for 15 minutes at 15 pounds pressure. Tweezers and razor blades may be sterilized in the cooker at the same time; wrap them in aluminum foil before placing in the cooker. At the same time sterilize pint jars of water to use for explant cleaning. Dr. Bridgen successfully propagated Boston fern rhizome tips, Africon violet leaf/petiole sections, and Wandering Jew shoot tips with this medium. Careful heed to cleanliness and sterile technique must, of course, be observed to avoid growth of molds or other contaminants.

For the fun of it try other explants or ingredients. For example, what happens if you substitute coconut milk for some of the tap water? (See African violet, p. 133, or Boston fern, p. 105). Ref. #26.

MEDIA PREPARATION

This section is written to help growers who are unfamiliar with general laboratory procedures, particularly those who have decided to take a do-it-yourself approach to media making as opposed to purchasing complete premixes. The section addresses concepts in the use of equipment and explains some fundamental chemical principles pertinent to media preparation.

It is important to consistently use caution, care and common sense. Work slowly at first in order to anticipate the consequences of your actions, to learn what to expect, and become familiar with lab language, concepts, and culture characteristics.

If you have elected to mix your own media "from scratch" then you will need to have some of the following chemicals on hand, depending upon what species you plan to tissue culture. If you are not certain of the cultivars, or how many cultivars you plan to try, then this list is for you because it covers a wide range of formulas.

Suggested beginning chemical inventory

Inorganic chemicals

Compound	*Formula*	*Amount to Purchase*	*Price (1986)*
Ammonium nitrate[1]	NH_4NO_3	500 g	$11.00
Boric acid[2]	H_3BO_3	100 g	3.70
Calcium chloride[1]	$CaCl_2 \cdot 2H_2O$	500 g	9.50
Calcium nitrate[1]	$Ca(NO_3)_2 \cdot 4H_2O$	500 g	9.20
Cobalt chloride[2]	$CoCl_2 \cdot 6H_2O$	25 g	8.40
Cupric sulfate[2]	$CuSO_4 \cdot 5H_2O$	250 g	6.50
Ferrous sulfate[1]	$FeSO_4 \cdot 7H_2O$	250 g	6.25
Hydrochloric Acid[3,4]	HCl (1 N)	1 L	5.00
Magnesium sulfate[1,5]	$MgSO_4 \cdot 7H_2O$	500 g	9.20
Manganese sulfate[2]	$MnSO_4 \cdot H_2O$	100 g	3.60
Potassium chloride[1]	KCl	250 g	4.50
Potassium iodide[2]	KI	100 g	10.30
Potassium nitrate[1]	KNO_3	500 g	11.10
Monopotassium phosphate[1]	KH_2PO_4	100 g	5.50
Sodium hydroxide[3,6]	NaOH (1 N)	1 L	4.50
Sodium molybdate[2]	$Na_2MoO_4 \cdot 2H_2O$	100 g	10.00
Monosodium phosphate[1,15]	$NaH_2PO_4 \cdot H_2O$	250 g	10.70

Compound	Formula	Amount to Purchase	Price (1986)
Sodium sulfate[1,15]	Na_2SO_4	500 g	9.00
Zinc sulfate[2]	$ZnSO_4 \cdot 7H_2O$	100 g	5.00

Organic chemicals

Compound	Common name	Amount to Purchase	Price (1986)
Adenine sulfate, dihydrate[7]	$AdSO_4$	5 g	6.85
Agar	Agar	1 kg	38.60
OR agar substitute	Gelrite	1 kg	65.00
L-Ascorbic acid[11,15]	Vitamin C	25 g	3.40
N6-Benzyladenine (6-Benzylaminopurine)[7]	BA (BAP)	100 gm	3.50
d-Biotin[9,15]	Vitamin H	100 mg	3.75
Calcium hypochlorite[12]	$Ca(OCl)_2$	250 g	7.30
Casein hydrolysate[14,15]		100 g	5.35
Charcoal, activated, neutralized[13]		500 g	7.00
Choline chloride[9,15]	Choline	100 g	3.95
Citric acid (anhydrous)[11,15]		100 g	4.20
Coconut milk[14,15]		1 fruit	1.00
L-cysteine HCl[15,16]	Cysteine	5 g	3.15
2,4-dichlorophenoxyacetic acid[7,15]	2,4D	100 g	5.95
Ethylenediamine tetraacetic acid disodium salt[8]	EDTA or Na_2EDTA	100 g	7.70
Folic acid[9,15]		1 g	3.20
Gibberellic acid[7,15]	GA_3	500 mg	8.35
Glycine[15,16]		100 g	3.75
Indole-3-acetic acid[7]	IAA	5 g	5.25
Indole-3-butyric acid[7]	IBA	1 g	3.40
myo-Inositol	Inositol	50 g	6.10
N6-(2-Isopentenyl)-adenine (6-(y,y-Dimethylallylamino) purine)[7]	2iP	100 g	6.00
Kinetin (6-Furfurylaminopurine)[7]	Kinetin	100 mg	4.50
1-Naphthaleneacetic acid[7]	NAA	25 g	3.40
Nicotinic acid[9]	Niacin	100 g	4.00
D-Ca-Panthothenic acid[9,15]	Pantothenate	5 g	3.15
Polyoxyethylenesorbitan[12]	Tween 20	100 ml	4.25
Pyridoxine HCl[9]	Vitamin B_6	5 g	4.00
Riboflavin[9]	Vitamin B_2	5 g	3.15
Sodium hypochlorite[12] (5%)	Clorox, Purex	1 gal	.80
Sucrose[10]	Sugar	10 lb	4.00
Thiamine hydrochloride[9]	Vitamin B_1	5 g	3.15
Yeast extract[14,15]		250 g	5.25

1. Major salt.
2. Minor salt.
3. Use to adjust pH.
4. Use to dissolve cytokinins.
5. Purchase Epsom salts from local pharmacy.
6. Use to dissolve auxins.
7. Growth regulator.
8. Iron chelating agent.
9. Vitamin.
10. Purchase sugar from grocery store.
11. For antioxidant solution to help prevent explant browning.
12. For cleaning explants.
13. For rooting media.
14. Undefined nutrient source.
15. Limited use in media, confirm need before purchase.
16. Amino acid.

Stock solutions

Stock solutions are concentrated solutions of single or groups of chemicals, portions of which will subsequently be combined to make the final medium. It is customary to prepare these stock solutions ahead of time in relatively large batches thus saving the labor and time involved if one were to weigh every chemical every time a batch of medium is made. Further, because the chemicals are weighed in multiples of requirements for any single batch of medium, greater ease in weighing and greater accuracy of quantities is achieved because small inaccuracies are lost in the larger weighings.

Stock solutions are stored in the refrigerator. There is little agreement on how long they may be held without losing some of their qualities. Inorganic stocks are more stable than organic stocks. A conservative practice is to store auxins and cytokinins for no more than two weeks, and all other stocks for no more than two months. If sediment or contaminants appear, the solutions should be discarded immediately.

Stock solutions should be made in such a way as to not form precipitates. Precipitates are insoluble compounds that separate out from a solution, usually appearing as a sediment in the bottom of the container. If any solution is used from a stock solution containing precipitates it will not possess the correct chemical balance because some elements will be tied up in the precipitate. None of the precipitated chemicals can be used by the plants. In order to avoid the formation of precipitates when making stock solutions one of two alternative strategies can be followed: (1) combine only those compounds which do not form precipitates at high concentrations, or (2) make the stock solutions weak enough that precipitates will not form.

By way of example, the following stock solutions are used in making a medium for blackberry culture. They should be trouble-free when using only purified water (deionized or distilled water) and carefully mixed.

STOCK SOLUTIONS FOR BLACKBERRY MEDIUM

1. *Major Salts*

Pour about 700 ml water into a liter flask.

Weigh the following chemicals and add to the flask while stirring with a magnetic stirrer:

	Grams (g)	
Potassium nitrate (KNO_3)	19.0	
Ammonium nitrate (NH_4NO_3)	16.5	
Calcium chloride ($CaCl_2 \cdot 2H_2O$)	4.4	
Magnesium sulfate ($MgSO_4 \cdot 7H_2O$)	3.7	
Monopotassium phosphate (KH_2PO_4)	1.7	
		45.3 g

After adding and dissolving all of the above, add water to the 1000 ml mark.

This stock solution is 10 times the formula concentration; 100 ml of this stock will be used in making a liter of blackberry medium. It will be noticed that this solution is weaker than most other stocks; this is because if it were any stronger it would tend to form precipitates.

2. *Minor Salts*

Pour about 700 ml water into a liter flask.

Weigh the following minor elements and add to the flask while stirring with the magnetic stirrer:

	Milligrams (mg)	
Manganese sulfate ($MnSO_4 \cdot H_2O$)	1680	
Zinc sulfate ($ZnSO_4 \cdot 7H_2O$)	860	
Boric acid (H_3BO_3)	620	
Potassium iodide (KI)	83	
Sodium molybdate ($Na_2MoO_4 \cdot 2H_2O$)	25	
The two remaining compounds:		
Cobalt chloride ($CoCl_2 \cdot 6H_2O$)	2.5	
and		
Cupric sulfate ($CuSO_4 \cdot 5H_2O$)	2.5	
		3273 mg

are required at too small amounts to weigh accurately on many balances. Therefore, weigh 25 mg (a convenient amount) of $CoCl_2 \cdot 6H_2O$ and 25 mg $CuSO_4 \cdot 5H_2O$ and dissolve them in 100 ml water. Ten ml of this solution will contain the desired amount, 2.5 mg of each for the stock solution. Pipet 10 ml of this solution into the mixing flask, and save the balance of cobalt/copper solution for further use.

Add water to the 1000 ml mark of the mixing flask. This final solution is 100 times the formula concentration. Only 10 ml of this stock will be used in making a liter of blackberry medium.

3. *Iron stock*
   Pour about 700 ml water into a liter flask.
   Weigh and add the following chemicals. Stir until dissolved after each addition.

	Grams (g)	
Ferrous sulfate ($Fe_2SO_4 \cdot 7H_2O$)	2.78	
Na_2 EDTA	3.72	
		6.5 g

Add water to make 1000 ml.

If either of these compounds does not dissolve easily, briefly heat and stir on the hot plate/stirrer. The final solution is 100 times the formula concentration; only 10 ml of this stock will be used in making 1 liter of blackberry medium. Store in a brown bottle in the refrigerator because it is not stable in light.

4. *Vitamin stock*
   To 700 ml of water add the following, stirring to dissolve as each is added:

Inositol	10 grams
Thiamine	40 milligrams

Add water to the 1000 ml mark of the mixing flask. Ten ml of this solution will provide 100 mg of inositol and .4 mg of thiamine, the amounts required for the blackberry medium.

5. *Benzyladenine stock* (BA). Concentration = 0.1 mg/ml.
   Weigh 25 mg of BA and place in a 400 ml beaker.
   Add about 5 ml water.
   With a dropper add one molar hydrochloric acid (1M HCl, see page 39) dropwise while stirring with a spatula until the BA is dissolved.
   Add water to 250 ml. This provides a stock solution which contains 0.1 mg of BA per ml. The blackberry medium to be made requires 1.0 mg of BA, so 10 ml of this stock solution will be included in the liter of medium to provide 1.0 mg of BA.

6. *Indole-3-butyric acid stock* (IBA). (optional for multiplying) Concentration = 0.1 mg/ml.

Weigh 25 mg of IBA and place in a 400 ml beaker.

Add about 5 ml water.

With a dropper add one molar sodium hydroxide (1M NaOH, see page 00) dropwise while stirring with a spatula until the IBA is dissolved.

Add water to 250 ml. This provides a stock solution which contains 0.1 mg of IBA per ml. For blackberry multiplying medium 0.1 mg of IBA per liter is optional.

7. *Gibberellic acid stock* (GA_3). Concentration = 0.5 mg per ml.

Weigh 50 mg of GA_3 and place in a 400 ml beaker.

Add about 5 ml water.

With a dropper add one molar sodium hydroxide (1M NaOH, see page 39) dropwise while stirring with a spatula until the GA_3 is dissolved.

Add water to the 100 ml mark. This provides a stock solution which contains 0.5 mg of GA_3 per ml. The blackberry medium to be made requires 0.5 mg GA_3, therefore 1 ml of this solution will be added to the medium.

SIMPLE PROPORTION

When you want to know the amount of stock solution to add to a medium, to provide a specified number of milligrams, the easiest arithmetic to apply is "simple proportion."

For example, you have used a total of 3273 mg of MS minor salts to make up a liter of stock solution. You need 32.7 mg in a particular formula so you need to know how many milliliters of stock solution to use.

Write down the original milligrams of salts and the original milliliters of stock solution. (It doesn't matter if you have used some, the *proportion* is still the same.) Then write the milligrams you need over the unknown milliliters.

$$\frac{3270 \text{ mg}}{1000 \text{ ml}} = \frac{32.7 \text{ mg}}{? \text{ ml}}$$

Cross multiply:

$$\frac{3270}{1000} \times \frac{32.7}{?}$$

$$1000 \times 32.7 = 3270 \times ?$$

$$32700 = 3270 \times ?$$

Divide both sides by 3270 so the unknown will be by itself.

$$\frac{32700}{3270} = \frac{\cancel{3270} \times ?}{\cancel{3270}}$$

$$\frac{32700}{3270} = ?$$

10 ml = the amount of stock you will need.

DISCUSSION

Having made the stock solutions, the biggest chore in mixing media is already done. The foregoing stock solutions will be sufficient for 10 liters of media, or more, depending upon their respective concentrations and the formula requirements. The individual stocks can also be used in other formulas as well, as detailed in Section II, in which case they will be used in combination with other stock solutions not detailed here but based on these examples. A case in point are the hormones. Other cytokinins besides BA (namely 2iP and kinetin) are mixed in the same way as described for BA. Other auxins (NAA and IAA) are mixed the same way as described for IBA.

The GA_3 could be mixed with the IBA to form a single stock solution (the quantities would have to be different), but that would preclude using them individually in other media formulas that do not require both of them.

The reason the BA and IBA stock solutions are made up by using 25 mg in 250 ml water is because 25 mg is the smallest amount that can be conveniently and accurately weighed on an inexpensive balance. As mentioned before, the hormones have a relatively short shelf life (they are not as stable as the inorganic salts), so ideally mix only as much as can be used in about two weeks, but that might be less than can be conveniently weighed.

There is not total agreement among labs as to how stock solutions should be stored or for exactly how long. If they have a short shelf life, as do organics, then they will be stable for a longer period if stored in the refrigerator. If they have a longer shelf life, as do the inorganic salts, then they can readily be stored in a lab cupboard but run a greater risk of growing contaminants because of the warmer temperature there. On the other hand, if the salt solutions are bordering on forming precipitates they will certainly do so in the cold of the refrigerator because the solubility of most compounds decreases with the temperature. Should they form precipitates they can be brought up to room temperature, or heated further, and used providing the precipitate redissolves.

It should be noted that the major and minor salts in the blackberry formula cited are the same as MS salts. This fact is significant, because MS salts are so widely used in standard formulas.

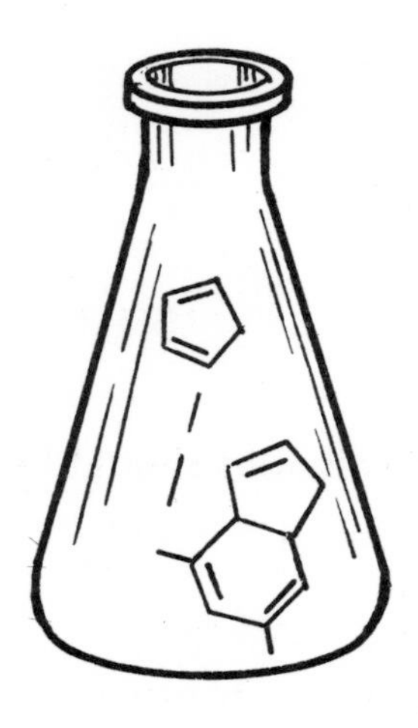

Processing Media

The time has now arrived to mix one liter of medium. The formula selected is a blackberry medium for which stock solutions were prepared as an example in the previous section. The complete formula is as follows:

Blackberry medium

Compound	*Milligrams*
NH_4NO_3	1,650
KNO_3	1,900
$CaCl_2 \cdot 2H_2O$	440
$MgSO_4 \cdot 7H_2O$	370
KH_2PO_4	170
$MnSO_4 \cdot H_2O$	16.8
$ZnSO_4 \cdot 7H_2O$	8.6
H_3BO_3	6.2
KI	0.83
$Na_2MoO_4 \cdot 2H_2O$	0.25
$CoCl_2 \cdot 6H_2O$	0.025
$CuSO_4 \cdot 5H_2O$	0.025
$FeSO_4 \cdot 7H_2O$	27.8
Na_2EDTA	37.2
Inositol (myo-inositol)	100
Thiamine HCl	0.4
N6-Benzyladenine (BA)	1.0
Indole-3-butyric acid (IBA) (optional)	0.1
Gibberellic acid (GA_3)	0.5
Sucrose	30,000
Agar	5,000

Adjust the pH to 5.2

References: #113, #176.

SUPPLIES

The following materials and solutions should be ready:

- Hot plate/stirrer
- pH meter
- Balance
- Pressure cooker and basket,[1] or autoclave
- 1 2-liter Erlenmeyer flask
- 1 10-ml graduated cylinder
- 1 100-ml graduated cylinder
- 2 100-ml beakers
- 2 10-ml pipets

2 1-ml pipets
2 medicine droppers
1 spatula
Wash bottle containing distilled water
Pitcher, or other dispenser
Weighing papers
65 test tubes and caps[2]
Distilled water
Stock solutions:
- Major salts
- Minor salts
- Vitamins
- Iron
- BA
- IBA
- GA_3

Sugar
Agar
1 molar NaOH and 1 molar HCl for pH adjustment

1. Refer to section on equipment, p. 27, for construction of a wire basket for the pressure cooker. Also see Figure 11-a.
2. Test tube closure such as KimKaps do not offer maximum protection against contaminants. Caps of this type should be lined with a piece of nonabsorbent cotton as illustrated (Figure 9-f). (Magenta Corp. sells a cap which has a baffle and filter ring which offers ideal protection against contamination.)

When mixing media or stock solutions, a running check sheet should be followed to be sure the right chemicals and amounts are added. If they are checked off the moment they are added there will be no doubts should the technician be interrupted. Following is a suggested form:

Table 2.

Name of formula: Blackberry multiplying

Ref.: Zimmerman and Broome, 1980 *Proc. Conf. on Fruit Plants,* USDA

Chemical	*Amt/L medium*	*Date (1982)* 7–21	7–31	8–22
Major salts stock	100 ml	✓	✓	✓
Minor salts stock	10 ml	✓	✓	✓
Vitamin stock	10 ml	✓	omit	✓
Iron stock	10 ml	✓	✓	✓
BA stock	10 ml	✓	✓	✓
IBA stock	1 ml	✓	omit	omit
GA_3 stock	1 ml	✓	✓	omit
Sucrose	30 g	✓	✓	✓
Agar	5 g	✓	✓	6
pH 5.2		✓	✓	✓

Procedure:

If stock solutions have already been prepared as directed in the previous section the formula to be mixed becomes far less formidable because it simply amounts to combining portions of the seven stock solutions and making two weighings.

–Place a 2-liter Erlenmeyer flask containing 600 ml distilled water on the hot plate/stirrer. (Always use a container with ample room for the solution to boil. For example, use a 2-liter flask for one liter of medium or a 4-liter flask for 3 liters of medium.)
–Slide the stir-bar into the flask and turn on the stirrer.
–Weigh and add 30 grams of sugar.
–Add 100 ml of major salts stock solution. (Remember to measure to the bottom of the meniscus, the slight curvature of the liquid surface in the graduated cylinder.)
–Add 10 ml minor salts stock solution. (Whenever you use a pipet it is wise to use a bulb pipetter (see Supplies list, p. 34.) If a pipet is used as a straw, too often the result is a mouthful of untasty or toxic chemicals.) You may use a graduated cylinder.
–Add 10 ml of vitamin stock solution.
–Add 10 ml of iron stock solution.
–Add 10 ml of BA stock solution.
–Add 1 ml of IBA stock solution. (optional)
–Add 1 ml of GA_3 stock solution. Use a pipet.
–Turn off stirrer and add distilled water to one liter.
–Turn on stirrer and adjust pH of medium:
Calibrate the pH meter according to the manufacturer's instructions.
Using a washbottle with distilled water rinse off the pH meter probe.
Lower the probe into the medium as it is mixing.
Observe the pH reading. If the pH is below pH 5.2 (too acid), use a dropper to slowly add 1M NaOH, a drop at a time, until the pH meter reads pH 5.2; allow the medium to mix after each drop. If the pH is above 5.2 (too alkaline), then slowly add 1M HCl, a drop at a time, until pH 5.2 is reached.
–Weigh and add the agar. If using Gelrite adjust pH after adding it.
–Turn on the hot plate/stirrer heat. Continue to heat and stir until the medium boils vigorously but do not allow it to boil over.
–Dispense the medium, about 15 ml per test tube. A one-liter Nalgene pitcher will do very well to dispense the medium into test tubes in a small lab. When the lab has grown to the point that large quantities of medium are used, an automatic pipetter should be purchased.
Avoid spilling medium on the rims of the test tubes because the cotton filter in the cap will then stick to the tube rim and come out when transferring. A trace of vaseline on the underside of the pitcher lip will help prevent drip.
–Cap the test tubes.
–Place the test tubes in racks (Figure 10-a) in the autoclave, or—
–Place the test tubes in the pressure cooker wire basket (see Figure 11-a).
There will be too few to stand up properly; fill the gap with an empty beaker.
–Process the tubes for 15 minutes at 15 pounds pressure according to the instructions in the manufacturer's guide.
–After processing and when the cooker or autoclave pressure is back to zero, remove the basket with tubes and place on an angle so that the medium will solidify on a slant (Figure 11-a).
–When they have cooled, label and store the sterilized tubes of medium in a cool,

clean place. Store them in the box in which they were shipped, or in test tube racks.

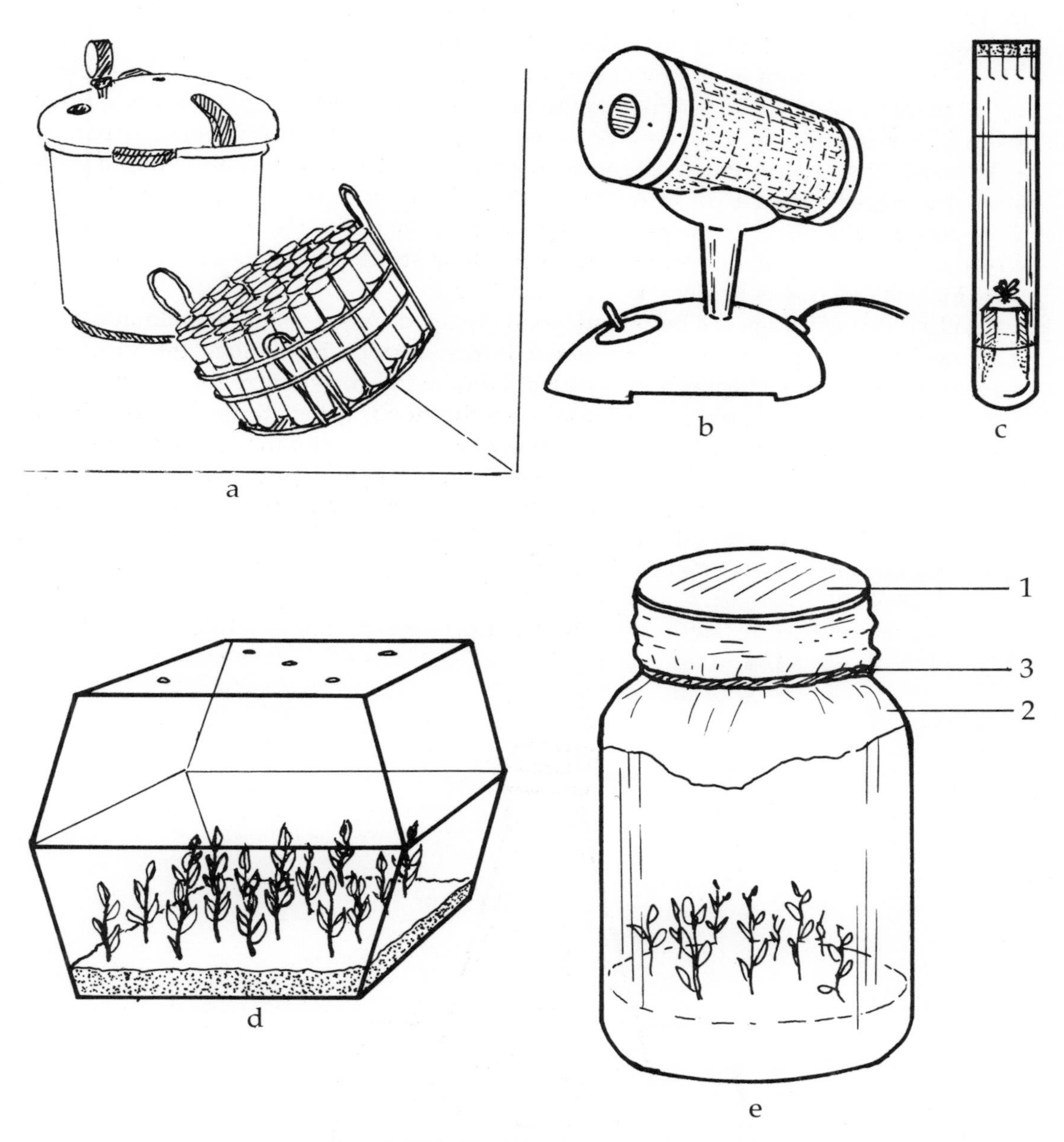

Figure 11. (a) Wire basket for holding test tubes in pressure cooker. (b) Bacticinerator, an alternative method for sterilizing instruments. (c) Test tube showing paper bridge. (d) Clear, plastic berry cups used for growing-on. (e) Pint jar showing (1) Polycarbonate, or glass, disc under (2) Plastic sandwich-wrap and (3) Rubber band.

Liquid media

Some standard formulas prescribe the use of a liquid medium for one or more stages of growth. Liquid media are faster to make than agar media because there is no agar which must reach the boiling point to dissolve before it can be evenly dispensed. In addition, changing to a liquid medium may be dictated by problems which the culture exhibits on a gelled medium. Such problems may include "bleeding," brown leaves, or poor growth.

Agitated liquid media will not allow build-up of waste products adjacent to the culture as can happen in gelled media. Some cultures in liquid do not require agitation (aeration) but, usually, such cultures are agitated on a rotator, shaker, or rocker. When cultures are started from inch long cuttings (2½ cm), or longer, the base of the explant may be placed in 3 to 5 ml of liquid medium in a test tube and no agitation is required. However, for smaller cultures, an alternative to agitation is to insert a filter paper, or paper towel, "bridge" in the test tube to support the culture while providing it with liquid. The paper acts as a wick carrying nutrient solution to the culture on top of the bridge (see Figure 11-c). Cut paper strips 3" × ¾" to insert in the test tubes. Place the two ends down so that the middle of the strip will be above the liquid and hold the culture. This will, of course, be done prior to sterilizing.

Another type of bridge can be assembled in the hood. Tubes of liquid medium and 3" × 3" squares of paper toweling are separately sterilized. To sterilize the pieces of toweling place them in a beaker, cover with aluminum foil and sterilize for 45 minutes at 15 pounds pressure in cooker or autoclave. In the hood, using sterile forceps, firmly grasp a piece of the toweling holding it between the forceps with the tips of the forceps at the center of the paper. With a twist of the wrist pull the paper into the test tube. If it is not grasped firmly the forceps will simply make a hole in the paper and slide through. When the paper is in the liquid on the bottom of the test tube, fold over the corners to make a platform on which to place the culture.

References: #26, #37, #62, #93, #105, #111, #113, #127, #130, #133, #141, #153, #156, #162, #164, #165

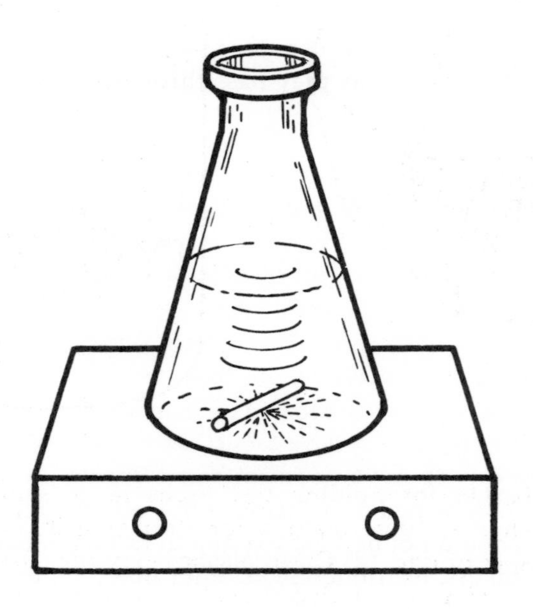

Chapter IV: (Micro) propagation— What It's All About

In this chapter you will, at long last, see the pay-off for all the effort and investment called for in the previous chapters. It is in this chapter you, as a grower, will return to the mainstream of growing. Even though it is a nursery in miniature you will soon feel at home, joyously heralding the healthy plants which multiply at incredible rates, and puzzling and worrying over those that do not. You will learn to detect subtle differences even though most, within a species, all look alike. You will learn to recognize stress and disease symptoms not now familiar to you. You will learn to recognize pinpoints of contaminating molds, or wisps of bacteria in the medium. You must keep reminding yourself that the air is full of contaminants and that much greater care than you use in your greenhouse must be used to keep them out of the cultures. But you are familiar with the whims of plants and you know the rewards of a good crop. You know also that persistence and patience usually tip the balance toward the successes so that the failures seem less important.

THE TRANSFER ROOM

The transfer room shares equal importance with the media preparation room. It is a room where cleanliness is of critical importance. The hub of the transfer room is the laminar flow hood, which has been discussed (page 28). It is in the transfer hood, or chamber, that the technician performs such important tasks as making thousands of transfers with speed and sterile technique, watching for plants which should be culled, making note of what works and what does not. In a small lab oftentimes it is up to the person doing the transferring to make important decisions as to size of transfer, which medium to use, and what medium adjustments to make, basing these decisions upon careful observations of the cultures as they pass through the chamber for division and fresh nutrients.

Sterile technique

An alert attitude in using sterile technique develops with practice; it is not easy to imagine and deal with the unseen. One must assume that everything—clothing, skin, countertops, instruments, air—that has not recently been sterilized, or sterilized and subsequently been protected, is teaming with mold spores, bacteria, and a host of other invisible enemies ready to invade clean cultures. One must "think" as bacteria to imagine where they are, or, more appropriately, where they are not.

The sterile air gently streaming through the laminar flow hood (Figures 8-1, 8-2) allows the technician to freely open culture tubes and make sterile transfers with reasonable assurance the cultures will not be contaminated during the operations. Except for the filter, the hood should be wiped down daily with a disinfectant (Zepharin, Physan, or Lysol). The HEPA filter should not be touched due to its fragile nature, and care must be exercised not to push beakers, instruments, or other objects into it.

Traditionally, instruments have been sterilized by dipping them in alcohol followed by flaming (burning off the alcohol) using a Bunsen burner or alcohol lamp. Alcohol used for disinfecting instruments should be 95% ethyl or isopropyl alcohol, because at 95% it will burn off more easily than will a more dilute form. There is a special Bunsen burner that saves gas and minimizes the danger of an open flame, a Touch-o-matic; a pinpoint pilot flame burns continuously to light the higher flame which is triggered by resting one's hand on a disc while flaming an instrument.

A convenient alternative to sterilizing instruments with alcohol is to use two solutions of household bleach (Clorox, Purex, etc., which is 5% sodium hypochlorite):

1/10 bleach: Measure 100 ml household bleach into a liter flask. Add water to the liter mark.

1/100 bleach: Measure 100 ml of 1/10 bleach into a liter flask. Add water to the liter mark.

Sterilize instruments by dipping them in 1/10 bleach for a few seconds followed by a rinse in 1/100. Instruments must be stainless steel or they will rust very rapidly in bleach. When using bleach, wear gloves because it will burn your skin; it will also stain or eat some fabrics.

Another option is an electrical sterilizer such as a Bacti-Cinerator II Sterilizer (see Figure 11-b). This little electric device sterilizes instruments in 5 seconds at 1600°F as the instruments are inserted into the red-hot, hollow cone of the cylinder. Two sets of instruments must be used alternately to allow one set to cool.

All three methods are used commercially and each has its drawbacks. The alcohol poses a fire hazard, bleach must be mixed fresh once or twice a day, and the electric sterilizer heats the instruments so hot it takes them a long time to cool. The decision as to which is the best method is a matter of preference. Many commercial labs use Bacti-Cinerators. A hot bead sterilizer is the newest option.

A few pointers to remember are: Avoid any obstruction to the laminar air flow because it will change the air flow pattern and so invite contamination. Do not have items such as beakers or racks of tubes between the HEPA filter and exposed cultures.

–Work at arms length, as far back in the hood as is practical.

–Do not make wide, sweeping arm movements over the work area such that contaminants will drop from sleeves or arms. Keep the right hand on the right hand side and the left hand on the left-hand side of the transfer hood work area. If crossing over is necessary use great care, keeping elbows as close to the body as possible.

–The hand that holds a test tube is contaminated by it and can contaminate forceps which, if not sterilized, can contaminate the inside of a tube when performing a transfer.

The transfer process

Having considered general matters of sterile technique, we can now address the transfer process itself. While out of order in terms of the total process from an explant to plantlets ready to grow on, the technical operations involved in dividing and transferring plant material, that is already started and multiplying, demonstrate the set pattern for all operations in the hood. You will need to be familiar with these operations, known as sterile technique, before you can successfully start an explant and establish it in sterile culture.

Equipment and supplies to have ready are (Figure 8-2):

–Bleach solutions (1/10 and 1/100) in plastic dishes (see Supplies list, page 31). Mix fresh solutions every 4 hours (see page 60 for mixing) because chlorine is released from open containers. Disinfect plastic dishes with Lysol and rinse with distilled water before pouring in bleach solutions.

–Stainless steel forceps, 10″.

–Stainless steel knife handle and disposable blade, or paring knife.

–Sterilized, commercial grade paper towels. Fold and place about 20 in a 600-ml beaker. Cover with aluminum foil. Sterilize 45 minutes at 15 lb pressure.

–Container for used caps. Place on the floor in front of the hood, beside the technician.

–Test tube rack (rack for 40 tubes, Figure 10-a) to hold used test tubes—on counter beside technician.

–Waste basket for used paper towels—on floor beside technician.

–Household plastic gloves.

–Sterile medium in tubes (see about jars, page 33).

–Tubes containing cultures ready to transfer, or new explants (preparation of explants will be described in the next section).

–Test tube racks (for 10 tubes) to hold new transfers for labeling (see Figure 10-b).

Preparation of the hood

–Wipe or spray hood with Lysol or equivalent, but be careful to avoid the HEPA filter.

–Turn on hood blower for 10 minutes before transferring.

–Disinfect hood counter with 1/10 bleach.

Transfer procedures

–Rinse gloved hands in bleach.

–Immerse knife and forceps in 1/10 bleach for at least one minute and rinse in 1/100 bleach. Drain.

–Using forceps and knife, lay a sterile paper towel on hood counter as far back as is practical to work.

–Return knife to 1/10 bleach.

–Grasp test tube containing culture in left hand near base of tube. Check identification label. Also check the tube for contamination.

–Holding sterile forceps in right hand, grasp test tube cap with right hand little finger. Twist slightly (to help keep cotton in the cap) and remove cap from tube.

–Still holding forceps and without contaminating them, allow the cap to drop into the cap container on the floor.

–Using forceps, remove culture from tube and place it on the sterile paper.

–Put the used tube aside in a test tube rack (for 40 tubes. When full the rack can go directly into the dishwasher). (See Figure 10-a)

–Pass the forceps to the left hand and pick up the knife.

–Dip knife in 1/100 bleach. Drain the knife.

–Using forceps and knife, trim and divide culture. Trim away any brown or dead material. Cut and divide appropriately.

–Dip forceps in bleach—1/10 then 1/100—and drain.

–With left hand, obtain fresh test tube containing medium.

–Still holding forceps, remove cap as described above.

–Still holding cap and forceps, use forceps to obtain a piece of plant material.
–Insert material into test tube. If placing on agar, be sure there is good contact between agar and culture.
–Replace cap on tube and place tube in test tube rack (for 10 tubes), ready for labeling.
–Return forceps to 1/10 bleach.
–After doing about 6 tubes, dispose of the paper towel with discarded material on it into the waste basket and obtain a fresh towel.
–When finished transferring all the tubes of one variety, label with marking pen, grocery store labeler, or hand stamp.
–Place tubes in Todd Planter trays (or Speedling trays) and move the trays to the culture growing room shelves (see Figure 10-c).
–Record the date, how many tubes completed, multiplication rate, medium used, any experiments underway, and when next transfer is due.

The foregoing procedures provide the guidelines for all transfer operations in a laminar flow hood. Other disinfecting systems will require only slight procedural changes that will be evident to the technician.

CLEAN UP

Tissue culture operations produce a constant flow of used containers that require washing for reuse. Some small labs have household dishwashers. The tines on the bottom shelf are bent down to accommodate racks of tubes to be inverted and placed on the shelves for washing. It is not necessary to remove agar from the tubes before washing. The hot water in the washer will dissolve and wash it away with no problem. A hardware cloth (wire mesh) cover is placed over the tubes before they are inverted to keep them in the rack.

Tube caps with cotton filters are seldom washed because they are resterilized with each new batch of medium. When they appear dirty the cotton is removed, the caps washed, and new cotton put in.

Jars are easily washed in the dishwasher. Usually the agar is removed first for more satisfactory cleaning. It should be disposed of promptly so that it will not grow microorganisms. It can be melted and flushed down the drain with lots of hot water, removed with other trash for commercial pick-up, or buried.

Culture media are an ideal nutrient for many contaminants. When cultures become visibly contaminated the molds, yeasts, or bacteria will quickly overgrow and kill the plantlets in culture. Occasionally, containers with media will become visibly contaminated before they are used for transfers. Regardless of when or how, whenever contaminants have multiplied so they form patches of growth on or in agar (or liquid medium is no longer clear) the billions of spores present pose a significant hazard of spreading in the lab. (It is doubtful if any of them are human pathogens, but there is always this risk. Most human pathogens thrive in a higher pH.) The safest means of disposing of these colonies is to place the contaminated containers in a pressure cooker or autoclave and sterilize before opening, emptying, and washing them.

EXPLANTS AND THEIR PREPARATION

An explant is a piece of a plant from which a culture is started. If all goes well, just one explant will produce thousands of plants. Of course, it is well to have more than one because of losses, but often only a few explants are necessary and one stock plant is usually sufficient to supply all the starts required. Losses typically result from contaminants, disinfectants, or plant condition.

Explants range in size from a microscopic .1 mm to stem pieces 2" or more. Meristems, shoot tips, stems, anthers, flowers, leaves, embryos, hypocotyls, seeds, seedlings, rhizome tips, bulb scales, flower buds, corms, or roots can serve as explants or sources of explants. Coaxing explants to start is probably the process with the most variables in the inexact science of plant tissue culture. Explants are taken from every conceivable plant environment at any time of year and from plants in a variety of conditions. The optimum state for any one of these variables is extremely difficult to predict. In short, when the entire range of possibilities for each factor is considered, it is apparent that no precise instructions can be furnished to insure success.

After having been detached from the source plant, explants are submitted to cleaning treatments they may or may not survive. The odds of success are greatest if a few explants (10–20) are started at weekly or monthly intervals instead of gambling many valuable explants on only one set of conditions.

Cleaning explants

The most difficult part of establishing an explant is cleaning it successfully. Following are guidelines to make the job easier:

–Use new shoots when possible because they are cleaner than old wood, thus easier to clean.

–Plants under cover or shoots from forced, dormant branches are often cleaner than field plants.

–Plastic sandwich bags fastened over actively growing shoot tips on stock plants will help keep them clean for cutting several days or weeks later. This set-up should be shaded to prevent burning.

–The smaller the explant, the less contamination there is to remove, but the larger the explant, the more tissue there is present to help establish it in culture.

–The weakest bleach solution effective against contamination on any particular explants is the strength that should be used. Unfortunately, only experience will provide this information; explant material may remain contaminated due to solutions too weak to kill the contaminants. On the other hand, material can be burned if solutions are too strong.

Once again, limit your risks by treating several small groups in various ways.

Explants are disinfected with various concentrations of household bleach which is 5.0% sodium hypochlorite. The bleach solutions are used alone or in sequence with other disinfectants.

To make 100 ml of household bleach solution of each of the following commonly used concentrations:

½ mix 50 ml bleach with 50 ml water.
1/5 mix 20 ml bleach with 80 ml water.
1/10 mix 10 ml bleach with 90 ml water.
1/100 mix 10 ml of 1/10 bleach with 90 ml water.

Other disinfectants used to clean explants are:
Ethyl alcohol or isopropyl alcohol, 70%;
Calcium hypochlorite, 0.8% (see page 130).
Hydrogen peroxide (H_2O_2), 3%.
Mercuric chloride, 0.1%.
Benzalkonium chloride (Zephiran), 0.1%.

One example of routine treatment of explants follows. It is performed in a series of 8 beakers that have been rinsed in 1/10 bleach.

- –Stir explants on magnetic stir-plate for 10 mintues in 1/100 bleach to which has been added about 10 drops of liquid dishwashing detergent or 2 drops of Tween 20 per 100 ml solution.
- –Rinse in distilled water.
- –Dip in 70% alcohol for 10 seconds.
- –Rinse in distilled water.
- –Stir in 1/10 bleach plus 2 drops Tween 20 per 100 ml of solution, for 15 minutes, using a magnetic stirrer. Some labs use a vacuum pump (at 25 mm of mercury) to help penetration of the disinfectant into crevices. Other labs have found an ultrasonic cleaner useful in cleaning explants.
- –While still in 1/10 bleach, move to the transfer chamber.
- –Rinse in sterile water.
- –Soak in fresh sterile water for 5 minutes, rinse again.
- –Dip in antioxidant if used (see below).
- –Trim on sterile paper toweling.
- –Place in tubes of liquid or agar medium.

If the foregoing treatment does not kill the contaminants, which will be discovered after a few days in the culture growing room (because they will have had time to grow out and be visible), then it is necessary to (1) stir in 1/10 bleach for 5–15 minutes longer, (2) extend the alcohol dip to a minute or longer, or (3) soak for 1 to 20 minutes in one or more of the other disinfectants listed above. Do not mix disinfectants together in the same container. If any of the foregoing treatments kill the explants, then weaker solutions should be tried for the same or longer periods. To test the tolerance of a set of explants to 1/10 bleach, remove a few of the explants at 5, 10, and 15 minutes instead of running all of them 15 minutes.

If the explants turn brown during treatment, dip in anti-oxidant solution before trimming. An antioxidant solution is made by dissolving 100 mg ascorbic acid and 150 mg citric acid in a liter of water. Sterilize before using in the hood.

Meristemming

This term is often used in an incorrect sense to mean micropropagation. Literally, and correctly, however, it refers to the utilization of meristematic tissue for explant material, usually the microscopic meristematic dome and the accompanying pair of leaf primordia. Meristemming is frequently used to establish a virus-free clone of plants. The theory is that meristematic tissue grows faster than plant viruses can infect these new cells, so the smaller (younger) the meristem explant, the greater the probability of obtaining virus-free plants. Some government labs heat treat, meristem, and test for viruses (virus index) before issuing virus-free strawberry plants.

Strawberries are easy to meristem, compared with most plants, but quite typical of the operation so a good plant to use to learn the procedure. For these reasons it is described here (see Figure 12).

To meristem a strawberry plant, begin by removing two or three inches of young runner tip before the leaves have expanded. Because the meristems are within protective buds, where they are initially sterile, surface sterilizing need not be as rigorous as described above. Stir 10 minutes in 1/10 bleach followed by one or more minutes in two sterile water rinses. In the hood, place a sterile petri dish under the microscope. Meristems must be excised under a dissecting microscope, preferably with 30x magnification. Place a piece of sterile paper toweling (3″ × 3″) in the petri dish. Moisten it with sterile water. Place the disinfected runner tip on the paper in the petri dish.

You will need a 6″ knife with disposable blade and watchmaker's forceps. You will be searching for two or three buds. At some distance from the tip (an inch or more), you will notice a bulge in the stem, with or without an accompanying bract. Using sterile technique as described earlier, slice the stem with the knife just below the bulge (see Figure 12). Make another cut at the point of attachment of the outer layer. Make a vertical slit through the loosened outer layer and tear it away with the forceps. Examine the base to find the bud. If very young it may be difficult to detect, lying flat, close to the stem. If you cannot locate the bud at the base, then examine the base of the sheath you removed. Dissect the bud by slicing away very thin slices from the base up to the meristematic dome. The dome and one or two pairs of leaf primordia will appear clearer than surrounding tissues. Touch the knife tip to this bit of tissue. Moisture will hold it there until you place it in the test tube and release it by lightly cutting the blade tip into the agar. Cap the tube, then examine it while holding it up to the light. You will barely be able to see the microscopic explant.

To locate the next bud continue to slice away the stem upward toward the tip until you find another bulge. Again, slice through the stem at the point of attachment of the sheathing, outer layer. Find and dissect the bud as described above. The third bud is the apical bud and will be more difficult to find because of the older leaves surrounding it. Continue to slice upward on the stem. Often a slice will reveal the base of a bud, or the dome will actually fall out. Knowing precisely what to look for, and where, will come quickly with experience and be very rewarding.

Media for starting explants

There are many specific recommendations in the literature for starting explants of particular species. A number of these are excerpted in Section II of this book. Success can never be guaranteed but it is more likely for those plants which have been well researched and commercially produced in culture.

A couple of general guidelines can be followed to increase the probability of success in getting explants started. First, if no formula can be found for culturing the plant in hand, or the prescribed formula is ineffective, try various standard media (both liquid and agar) starting with those that have been used successfully for related plants (the same family or genus). Another useful approach is to start herbaceous perennials in half strength MS salts without hormones while woody plants should be tried in McCown's Woody Plant Medium (WPM) without hormones.

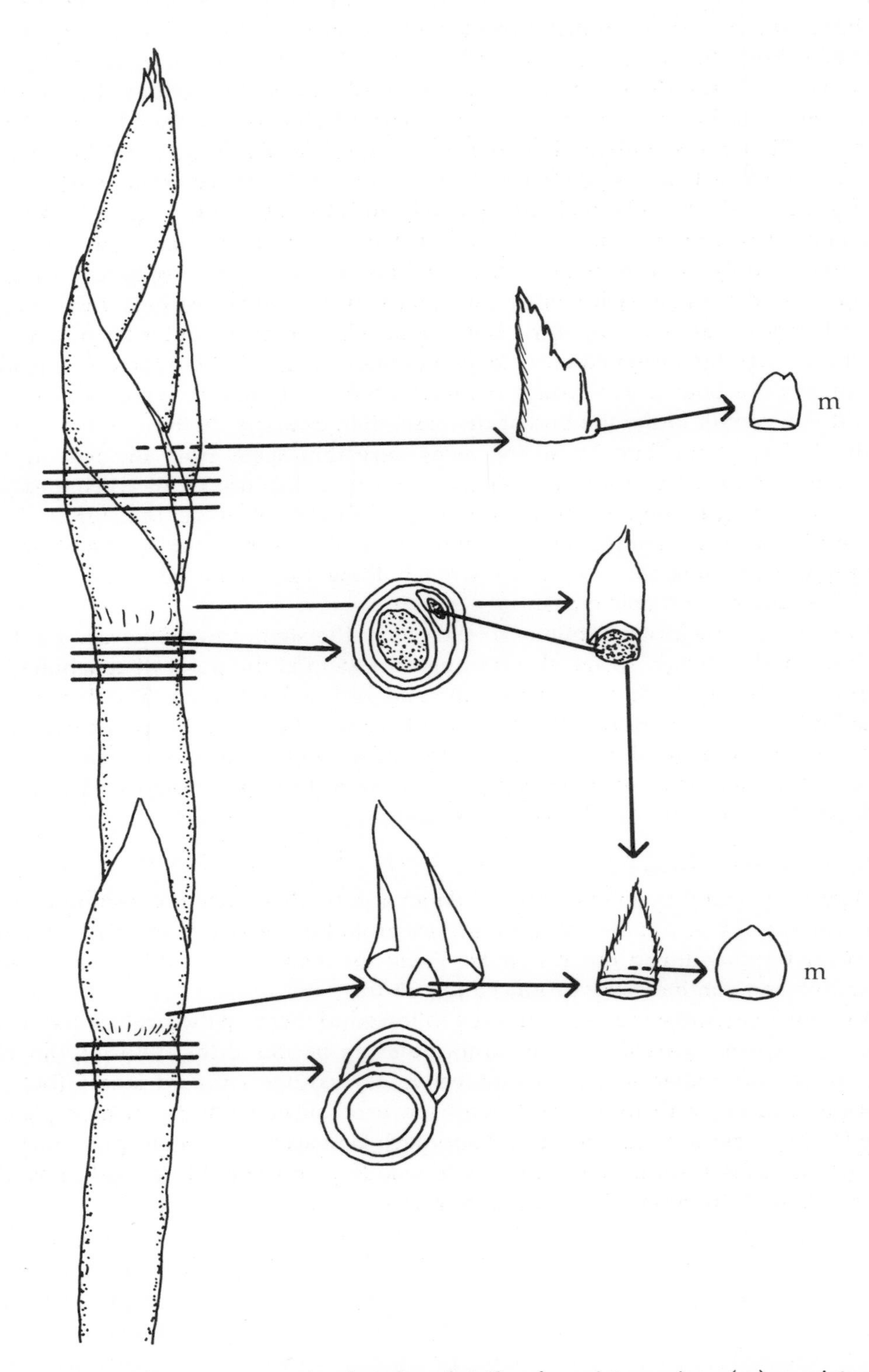

Figure 12. Strawberry runner tip showing details of meristemming. (m) meristem.

Culture manipulation

Once an explant is started the question is what to do next. Never open a culture tube, or other container, with a clean culture growing in it, except in the hood using sterile technique. Do not be in a hurry to divide it or to remove the new growth from the explant (see the transfer process, page 60). Often, with woody plants, it is wise to leave a portion of the original stem, attached to the new growth, and gradually remove it over two or three transfer cycles. A lateral shoot of a rhododendron explant, for example, should be an inch long before removing all of the original stem (Fig. 13).

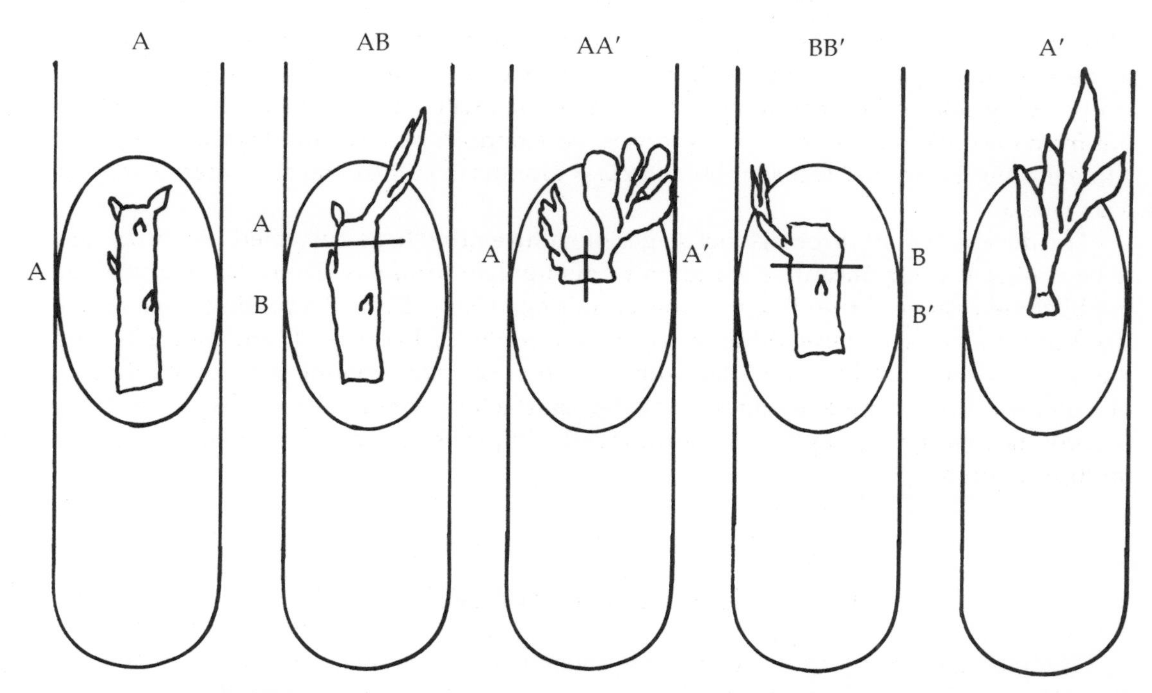

Figure 13. Rhododendron explant and some subsequent transfers. A Explant, AB Divide. AA′ Divide. BB′ Divide (discard B′ if dead). A′ Shoot removed from original explant.

Herbaceous material, on the other hand, will often involve a much smaller explant. The growth from a strawberry meristem, for example, may be readily visible within a week but may not be more than an eighth inch in a month. At this size (and/or time) it will benefit from "rolling it over." Simply move it (using forceps and sterile technique) to a new location *on the same agar slant,* which serves to place it in fresh medium and insure good contact of the new growth with the agar. In another one to three weeks it will have become a ¼-inch green ball or base, with many new shoots. Cut this in two and place each piece in a new tube.

A culture can be considered to be established and into the multiplication stage as soon as it is successfully divided. Then the question is how big should the transfers be and how often are they to be transferred? One rule of thumb that applies to size is to transfer a ¼-inch diameter piece when material has primarily basal growth (largely herbaceous species); or transfer inch-long pieces when cultures are primarily extended shoots and lateral breaks. In leafy material such as blackberries, where growth may be either from enlarged basal growth or extended, dichotomous growth, separate and transfer individual stems from the basal area and make nodal cuttings of the top material.

Often, shoots or stem pieces are laid down on the agar to provide maximum contact between the culture and the agar. However, if the leaves "bleed" where they rest on the agar it is better to place just the base of the stem piece into the agar.

The length of time between transfers will vary depending upon the variety, its culture history, size of transfer, media, and other factors. It is time to transfer when

the medium is discolored
leaves turn brown
growth slows or stops
tubes become crowded.

Ideally, cultures are transferred just prior to any of these events but that time is difficult to determine. Experience is probably the best teacher. Some cultivars require transfer every two weeks, others six weeks. The time cycles will vary not only with the species but from one stage to another. Strawberries, for example, may require transfer every 10–14 days after the second transfer, but after 2–3 months in culture the best interval may be four weeks.

The designation of three distinct stages of culture growth, as described by Murashige, is becoming less significant as we learn to manipulate formulas. There is, of course, an establishment stage (Stage I), a multiplication stage (Stage II), and a rooting stage (Stage III), but in most cases three different media are not used because (1) the multiplication medium serves as well for the establishment stage, (2) often rooting is done *in vivo* (out of culture), (3) a single medium may be satisfactory for all three stages, or (4) a prerooting medium (possibly no hormones to improve height) is applied in the late multiplication stage.

THE CULTURE GROWING ROOM

The media preparation room and the transfer room activities support the activity of the most important room, the culture growing room. Culture growth determines the success of the whole business; it is the reason for precise media making, painstaking transferring, and the basis for viable plants to sell. Knowledgeable tissue culturists spend a part of each day monitoring the cultures in the growing room. The person who monitors the cultures coordinates the media making and transfer room schedules to meet production requirements.

A typical culture growing room uses cool white fluorescent lamps controlled by a 24-hour timer to provide 100 to 300 foot candles of light to the cultures 16 hours a day. The timer is set so that the 8-hour dark period occurs during the daytime and the 16 hours of light fall mostly during the night. This arrangement uses heat given off by the lights at night to help warm the growth room when heat is most needed. The light cycle required by the cultures is not affected if growing room lights are turned on for intervals during the day to do chores.

The question is sometimes asked if Gro-Lux lights are better for tissue cultures than cool white fluorescent lights. Authorities disagree on the answer to this question. Because cultures have an abundance of sucrose, light may be a minor variable. Very little photosynthesis is taking place in cultures but the fact that they are usually green indicates the presence of chlorophyll synthesis. The spectral curve of light from Gro-Lux lamps coincides more accurately with the requirements of chlorophyll synthesis than cool white light; however, common practice and some comparison studies indicate cool white tubes adequate for most genera. Occasionally Gro-Lux lamps are specified in the literature in which case it may be safe to assume that the scientist doing the research had reason for his/her preference. (Fig. 14)

Light values specified in the literature may be presented in other than foot candles. Micro-Einsteins, lux, lumens and micromoles are other expressions and equating them may be inaccurate. One "equality" presented is: 107500 Lx = 2000 $\mu E/cm^2 sec^1$, (Doug Waldren, "Data Acquisition Manual: Remote Ecological Environments", 1985, Weyerhaeuser Co.). Other excerpts from the literature describe lighting from fluorescent tubes at 100 $\mu E\ m^{-2}s^{-s}$; 76 μ mol $s^{-1}m^{-2}$; and 3500 Lux; all of which presumably fall within the range of 100–300 foot candles.

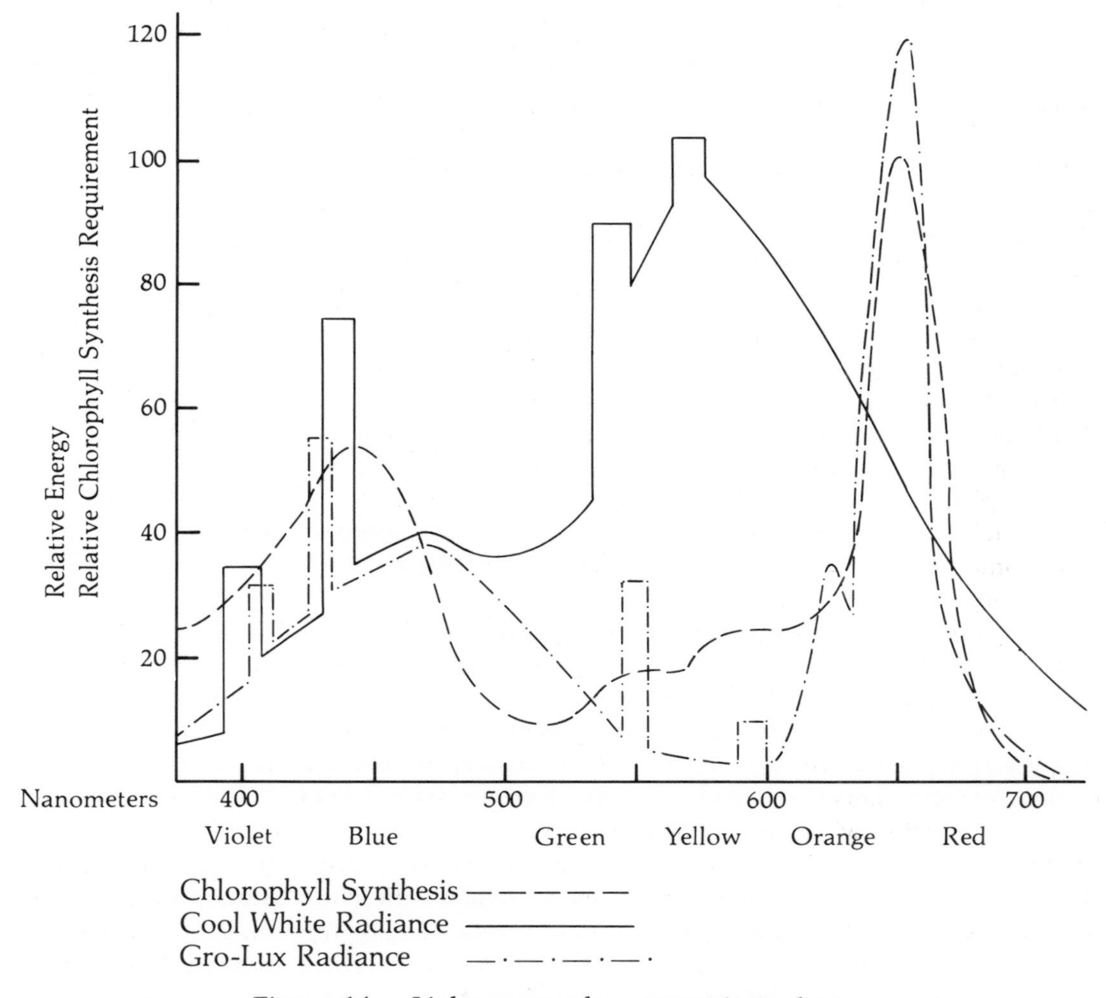

Figure 14. Light wave value comparison chart.

A centrally located themostat set at 77°F (25°C) controls heating and cooling equipment. Such equipment usually has a blower which is left on continuously for air circulation. Good air circulation is essential to minimize hot areas over lights and ballasts, or cold areas near the floor.

Most cultures do well under these conditions. There are exceptions, of course, cultures that do better in higher or lower light intensities, or even in darkness for one or more stages of culture. Amaryllis and orchids, for example, are often cultured in continuous light, while hosta can multiply in continuous darkness. Gerberas respond best to 1000 f.c. for the rooting stage compared to 100–300 f.c. for normal tissue culture growing.

Contaminants

Typically, the contamination rate in a production culture growing room can be expected to run about one percent. Obviously, as the contamination rate increases the cost-effectiveness of the operation decreases.

Cultures overrun by yeasts and molds are easily identified. Bacteria which grow as a cloud or haze within the medium, often surrounding the submerged base of the culture, are more difficult to see, especially if the agar is not perfectly clear. The easiest way to detect these contaminants is to examine the medium while holding the test tube up to a fluorescent light. This type of contaminant is insidious, it delays growth, impairs rooting and kills slowly. To date, the use of antibiotics and fungicides in media has been largely ineffective. Culture mites are another specter; they travel invisibly on air currents carrying with them bacteria and spores. The best course of action is to sterilize and discard all contaminated cultures immediately upon discovery. Next, sterile procedures must be reexamined to discover possible sources of contamination. All reasonable steps must be taken to prevent the spread of existing contaminants or the introduction of new ones.

Bacteria are best observed when Gelrite (a very clear agar substitute) is used because many of them grow below the agar or gel surface especially if they are emitted from within the stem through the lower cut end. Some headway has been made on the use of antibiotics to rid explants and culture of bacteria. One study used a combination of (per liter) 25 mg cefotaxim, 25 mg tetracycline, 6 mg rifampicin, and 6 mg polymyxin B. Because of chemical breakdown if autoclaved these substances must be cold filter sterilized before adding to sterilized media. They are also used as a soak in explant cleaning. While by no means a cure-all antibiotics are effective against at least some woody plant bacteria especially when used both as a presoak and in liquid media. Gentimicin, Streptomycin, and penicillin-streptomycin solution have all been credited with some success. Gentamicin and streptomycin appear useful in some instances even when they have been autoclaved.

Progress is also being made in antiviral substances. One example is the successful use of Ribavirin (Virazole) against several potato viruses. Another antibiotic discovery was the use of Agramine (methylolurea, UF-85) against certain yeasts in tulips and iris.

A special method of testing for the presence of contaminants, called indexing, is employed by some tissue culturists. Media that is more conducive to the growth of various contaminants than standard media is employed to optimize conditions in which the contamination will grow. By this method contaminants are forced to appear quickly so the culture can be discarded before unknowingly multiplying the contaminants. Tissue culture supply companies stock several media with which to perform these tests. The following indexing media are commonly used: (Bacto) nutrient broth, Sabouard's medium, yeast and dextrose agar, or potato dextrose agar. These come packaged, ready to use, with mixing instructions on the container. A small amount of culture material is minced and introduced into the medium. A few days in the culture growing room will grow at least some of the contaminants present (if, indeed, they are present) so they will be clearly visible. Indexing is particularly valuable to reveal organisms internal to new cultures which no amount of external treatment of the explant will eliminate.

Problem solving

In addition to contaminants, a careful monitor watches for "bleeding," phenolic exudates from the cultures that color the medium an inky-like purplish black. This condition has been blamed on dull knives, old cultures, or too liquid a medium. Actually, very little is known about "bleeding" except that it usually has an adverse effect on the culture, turning the base black, and retarding its growth and multiplication. It is worthwhile to retransfer "bleeding" cultures as soon as the symptom is observed.

Other culture conditions growers look for are changes in color or growth patterns. When leaves turn yellow or brown, or grow too fast with a watery succulence (vitrification); when stems turn red, or unwanted roots or callus are produced, it is time to vary the formula or the environment. When cultures produce roots the plantlets tend to mature and cease to multiply. If this happens in Stage II, production is sometimes curtailed and the hormones must be changed.

Changing the hormones is one of numerous formula modifications possible to try to remedy culture problems. Standard formulas do not always achieve the desired results for various, often unknown, reasons. Adjusting formulas to accommodate the special needs of cultures is sometimes necessary for their success.

Because of the dramatic influence hormones have on culture growth, the prime suspect to investigate when problems arise is the cytokinin/auxin ratio in a formula. A methodical way to establish the optimum cytokinin/auxin ratio is to draw a grid and plot the two factors in increasing increments on the two axes (see Table 3). For shoot cultures in the multiplication stage, rule out the ratios where the auxin content is greater than cytokinin content (at least in the first trials). Mix a medium of each extreme ratio, one or two other ratios, and the standard formula for comparison. Grow at least ten tubes in each variation, preferably through two transfer intervals. Sometimes the cultures grown in the extremes show dramatic differences, possibly a helpful clue. For example, a 0/0 ratio (no hormones) has been known to promote elongated shoots where short shoots had been the problem. No hormones was not the whole answer but it indicated that the problem was hormonal and that less cytokinin (or a different one) should be tested.

Table 3. An auxin and a cytokinin are plotted on the two axes in increasing amounts. Stage II cultures usually require cytokinin to be in excess of auxin. A preliminary test might include the six ratios underlined.

Cytokinin (mg/L)

Auxin (mg/L)	0	.5	1	3	5	10	
0	<u>0/0</u>	.5/0	1/10	3/0	5/0	10/0	
.5	0/.5	.5/.5	1/.5	<u>3/.5</u>	5/5	10/.5	
1	0/1	.5/1	<u>1/1</u>	3/5	5/1	10/1	
3	0/3	.5/3	1/3	3/3	<u>5/3</u>	10/3	
5	0/5	.5/5	1/5	3/5	5/5	10/5	
10	0/10	.5/10	1/10	3/10	5/10	<u>10/10</u>	

Other modifications to media, which can be explored as possible solutions to problems, need not be as involved as searching for the ideal cytokinin/auxin ratio. These can be straightforward, such as simply using ½ strength of the salts prescribed, or a different hormone or combination of hormones, pH, or agar concentration. Whenever a formula modification is used, always run a few controls or standard tubes at the same time for comparison. The logic underlying this approach is to grow a number of tubes with the medium that was presumably causing the problem so that a fair comparison can be made between the standard and the modified formula to evaluate any difference.

The number of possible media modifications is infinite. A few further variations and concentrations are suggested below. They should be tried singly or in combination and correlated with the symptoms which appear in the growing room (see page 74).

Salts concentration (¼, ½, or double strength)
Cytokinins—various concentrations, various cytokinins:
- 2iP—0 to 30 mg/liter
- Kinetin—0 to 10 mg/liter
- BA—0 to 10 mg/liter

Auxins—various concentrations (0–10 mg/liter) and various auxins (IBA, IAA, NAA)
- Vitamins—with or without, at standard amounts
- pH—various pH: 4.5–5.8

Agar—various concentrations: 0–10 g/liter, various name brands, sources and quality. Try Gelrite/Agar, 3/1 grams.
Air—allow more air to cultures by placing non-absorbent cotton in a hole in the polycarbonate lid before sterilizing the media, or use Magenta lids.

Refrigeration

There are, of course, many other problems besides that of determining the correct medium for a given plant species. For example, what happens when the transfer crew has not kept up with production and a thousand containers are behind schedule? A walk-in refrigerator will do much to solve such scheduling problems. Not only can it ease emergency situations in production but it can reduce greenhouse fuel bills by slowing the growth of plantlets ready for transfer to soil in midwinter until early spring. Some stock cultures can be held almost indefinitely in a refrigerator with subculturing done only once or twice a year. In such cases, the temperature is held at about 2°C (35°F) without light. Cultures that tolerate cold storage can be placed in it at any stage, not just the rooting stage. It is desirable to check plants weekly until confident of their storage life.

Cold storage is also useful to administer chilling treatments. Pear and apple cultivars respond with additional growth when chilled for 1000 hours prior to greenhouse planting.

Air

The need for air exchange in culture vessels is not a common problem. Most culture vessel systems appear to provide sufficient exchange for normal growth. However, in my lab potatoes exhibited spindly growth and tiny leaves in vessels that rarely caused problems with berries, rhododendrons, begonias, and numerous other plants.

An easy solution was to bore a ¼ inch hole in the polycarbonate disc covering the pint jar and install a piece of non-absorbent cotton in the hole. This small allowance for air exchange made a dramatic difference. Other covers that allow air exchange are Magenta-type baby food jar lids and their pint jar counterparts, Magenta GA7's that have a filter, and cotton stoppers as in flasks.

Other reported efforts to aerate cultures have involved pumping sterile air into the cultures. In one case open culture containers were placed in plastic bags supplied with pumped in sterile air. In another case an intricate system of tubing conveyed pumped, fresh, sterile air to each jar. The advantages claimed were lack of vitrification and greater ease of acclimatization.

According to the *Ball Redbook* tomatoes are indicator plants for ethylene. Their leaves bend down and the nodes are short, a condition called epinasty, when in its presence. We placed tomato seedling plants in with the potato cultures to test for ethylene. The results were mixed, therefore inconclusive.

Thomas and Murashige (1979) examined culture vessel air for 5 gases by gas chromatography. Compared with callus and nodal explants, CO_2 was lowest in vessels with proliferating shoots, which implied some photosynthetic activity. Ethane quantity was usually the same as in room air. Acetylene and ethanol were virtually absent from shoot cultures. Ethylene production was quite variable but usually higher in callus cultures.

Wilkins (1984) reports a long list of plant responses to ethylene. Some that could apply to shoot cultures in vitro are: seed and bud dormancy, inhibition of shoot elongation, promotion of radial growth, respiratory changes, leaf epinasty, root and root hair initiation, tissue proliferation, and apical dominance release. Under certain conditions most of these responses can also be induced by auxins. Auxins, CO_2, and stress stimulate ethylene production. Mele et al (1982) reported restricted growth of carnation in sealed vessels and in controls when ethylene was added.

The small leaves of potato cultures in sealed vessels may or may not be due to ethylene. Ethylene is suspect because the data indicate its prevalence and diverse influence.

By now it should be apparent to the reader that the state of the art of micropropagation is about where medicine was two hundred years ago. I have provided some guidelines that represent current practice, but the lack of specific information is both challenging and frustrating. From years of experience growers know how to deal with many of the requirements and idiosyncracies of particular cultivars growing in a greenhouse. Even so, plants do not always perform as expected or in the same way for one grower as for another. Significantly less experience has been gained with tissue culture practices so the grower must be prepared to manipulate media as problems arise.

As opposed to machinery (and even cars can be baffling) the ills confronted in cultures are comparable to human frailties. The variability is infinite. Often we do not know the causes or cures for problems. This is even more true in so young a science as plant tissue culture. When culture response is not as hoped, all we can do is try various alternatives based on what fragile information we have.

The following list of problems and possible solutions is admittedly inadequate because it does not say, "if you do A and B for C then D will happen." You, and hundreds of others in the next few decades, will fill in the blanks in this ever growing, ever challenging science.

Symptoms	*Possible causes*	*Possible answers*
Explant dies	Too harsh disinfectants.	Use weaker disinfectant.
	Media too strong	Use ½ or ¼ strength.
	Wrong stage of growth.	Obtain explants at different stage of growth.
Culture blackens and dies	Contaminated	Discard with care. Review sterile technique and sanitation.
	"Bleeding"	Transfer before it dies. Transfer more frequently. Use antioxidant in medium.
	Agar problem	Try different agar.
	Water problem	Check water purity.
	Wrong formula	Try different formula.
Explant live but no growth	Dormant	Chill for a month.
		Obtain explant at different stage of growth.
	Media too strong	Lower salts and hormones.
	Wrong formula	Try different formula.
Culture live but no growth	Too cold or hot	Change temperature.
	Wrong formula	Try different formula. Some stubborn explants have responded to a layering of media. Place explant (or subculture on agar slant, add 4 ml of the same medium without agar, thus submerging the explant. Add charcoal.
Growth too slow	Too cold or hot	Change temperature.
	Wrong medium	Try different medium.
Shoots too long (leggy), and poor multiplication	Too little cytokinin	Increase cytokinin. Increase air. Run cytokinin/auxin grid.
Shoots too short	Hormones too strong	Decrease or omit hormones.
No multiplication	Too little cytokinin	Increase cytokinin. Run cytokinin/auxin grid.
	Needs chilling	Cold store 4–8 weeks.
	Too cold	Increase temperature.
	Requires dormancy period	Cold treat 3–8 weeks.
Fat stems, small leaves, pale	Too much cytokinin	Decrease cytokinin.
Unwanted callus	Wrong hormones	Decrease or omit hormones. Run cytokinin/auxin grid. Try 1–5 mg/L TIBA (Triodobenzoic acid). Omit auxin.
Leaves chlorotic	Contaminant	Index for contaminants.
	Too hot	Decrease temperature.
	Wrong formula	Try different medium.

Symptoms	*Possible causes*	*Possible answers*
Leaves succulent (watery), "vitrification"	Osmotic potential upset	Decrease temperatures. Allow more air to culture. Increase agar strength.
	Too high cytokinin	Decrease hormones. Add phloroglucinol.
	Wrong agar	Try different agar.
	Culture too old	Transfer more often.
Premature rooting	Wrong hormone balance	Transfer more often. Increase cytokinin, decrease auxin.
Red stems	Stress	Change light and/or temperature.
	Too much sugar	Lower sugar content.
	Not enough NO_3	Increase nitrate.
	Culture too old	Transfer more frequently.

This list will trigger other ideas as to the causes of problems and their potential answers. For example, the influence of container size has long been a subject of controversy among growers. Obviously, it is easy to test the response of a culture to containers of various sizes. Other ideas will occur to the aggressive grower.

Recording the response of cultures to different media is of utmost importance. The results of every trial can provide important information upon which future operations must be based. The failure to maintain good records represents wasted present effort and lost future opportunity. It is important to note both good and poor growth since positive response proves both a favorable medium and good growing conditions which should be continued and built upon.

As problems arise you will want to:

–Examine lab records to learn when a problem started. If other culture varieties exhibit the same problem, check the medium history to learn what previous response has been. Check the original formula to be sure current ingredients agree or, if any changes were made, whether they were beneficial.

–Check all instruments to be sure they are functioning properly, particularly the pH meter, sterilizer, and water purifier.

–Try variations of medium, light, and temperature.

–Study the habits and requirements of the plant in question to see if these factors under normal greenhouse or field conditions yield any clues for tissue culture treatment.

–Consult outside sources of help including (see Appendix):

Professional organizations
Tissue culture supply companies
Cooperative Extension Service
Other tissue culture labs
Universities—botany and related departments, libraries

References: #37, #62, #82, #93, #95, #105, #146, #147, #162, #172, #173.

HARDENING-OFF

Tissue-cultured plantlets are often difficult to establish in greenhouse conditions. They behave as very tender seedlings so must be acclimatized with the utmost care. So much time has been spent on searching for desirable growth formulas to establish and multiply cultures that the important phase of acclimatization or hardening-off has been largely neglected and left to the hit-or-miss assumption, "that's the grower's problem." It is indeed, but obviously if growing-on techniques are neglected the time spent in multiplying can be largely wasted due to high losses in this later stage.

Electron microscopy has revealed the atypical leaf characteristics which usually develop in culture including stomata which do not close, excessive stomata per given surface area, and the lack of epicuticular wax. All of these anomalies account for greater water loss than is found in normally propagated plants grown in lower humidity.

To investigate this problem researchers in one instance reduced culture jar humidity using a layer of lanolin on the surface of the medium and a small bag of sterile silica gel hung in the jar. Humidity of 33% was achieved but, under circumstances of the experiment, this low percentage was detrimental to growth of the cultures. The best solution to the problem is to reduce the humidity for Stage IV plantlets so gradually that dehydration does not take place.

Beyond morphology and transpiration being disrupted, the entire photosynthetic process is confused in in vitro plantlets. Little use is made of CO_2 as the plantlets draw upon sucrose in the medium for energy instead. Again, the transition to greenhouse conditions must be gradual and with healthy, well grown plantlets.

The quality of plants grown-on depends on their condition in vitro. There is a trend to omit the rooting stage in culture and root plantlets in "soil" directly from Stage III. The success of this practice depends on the particular cultivar, its ease of rooting (in or out of culture), and on the grower. Oftentimes the roots that develop in Stage III are not functional in soil.

The advantages of skipping Stage III include less costly field labor, stronger root systems, reduction of transfer room use, and more space in the culture growing room. One company reported reducing the production cost of Rex begonias by 50% when they changed to direct rooting. But the plantlets of some species may need Stage III for more rapid rooting, better height, more sturdiness or simply conditioning.

In any case the quality of Stage IV plantlets depends on the quality of Stage II microcuttings or Stage III plantlets that go to Stage IV. Stage III medium often benefits from the addition of charcoal which appears to help remove residual cytokinin. Charcoal in the media is an excellent way to quickly distinguish rooting medium from multiplication medium, but it makes observation of root formation more difficult. Lower salts, especially nitrates, often helps induce rooting in Stage III media. The particular cytokinin used in Stage II has been reported to affect rooting. It was determined that liriope, schefflera, and philodendron survive transplant better by 80% when 2iP or kinetin are used in Stage II instead of BA. In Stage III cytokinin is usually eliminated and the auxin level raised. The auxin level is important; more is not better as high levels can be worse than none at all. Sometimes two auxins used together work better than either one alone. Sucrose levels are often lowered for rooting but occasionally they are raised. Phloroglucinol has been slightly beneficial in rooting of some fruits. Lighting requirements vary but many plantlets root better in increased light (350–600 f.c.). Some bulbs and fig root better in darkness as do some rosaceae for root initials. Agar stiffness is often increased for a drying effect but roots are more apt to break off when plantlets are removed. If liquid medium is used for Stage III many more root hairs will develop but roots may be less sturdy than with agar.

As plantlets are moved from Stage II or Stage III to soil mixes it is desirable to wash them to remove as much agar as possible because molds, yeasts, bacteria, and insects thrive on nutritious agar. Various fungicides should be tested on a few plantlets to learn which ones can be used safely. Boulay suggests 125 mg/L benlate. Do not use them unless it appears necessary. Some growers use a hormone powder or auxin dip at this point, but it is of questionable advantage considering the labor involved. Some growers remove the culture container lids to accustom the plantlets to lower humidity, but this practice is hazardous as cultures may dry out too quickly and become contaminated. Another approach is to transplant shoots to pots of soil mix which are placed in jars and covered as they had been in culture. Once rooted and established in the soil the humidity is lowered.

A California walnut grower has a patented process involving pretreatment in culture, root induction in flats in a grow room, and field establishment without a greenhouse stage. The root-ready plantlets in the field are first covered with plastic drinking cups which in turn are covered with 2 foam cups. The cups are removed in three stages. The method was used successfully with *Juglans paradox* and *Prunus persica.*

There must be almost as many soil mixes as there are growers. Soil media are not always defined in the literature which slights this very important stage. Some of those used for micropropagated plants include:

Jiffy 7's (peat pellets). Raspberries, carnations.
Peat moss, sand, and perlite, 3/1/4. Fruit tree rootstocks
Peat moss, perlite, sawdust. Rhododendrons
Peat moss, perlite, sawdust, pumice, vermiculite. Woody plants.
Peat moss, perlite, 1/2. Rhododendrons
Peat moss, perlite, 1/1. Rhododendrons, blackberries, apple, birch.
Peat moss, sand, 1/1. Grapes 1/1½ Loblolly pine
Peat moss, perlite, vermiculite. Acacia, 1/2/1 Larch, pine
Peat moss. Kalmia
Milled sphagnum moss. Top dressing for rhododendrons.
Peat moss, bark, sand, sawdust. Woody plants
Peat moss, vermiculite. Cattail, alder, willow
Vermiculite, sand 1½/1. Loblolly pine
Peat moss, pumice 1/1. *Pinus radiata*
"Mica-peat" + NAA solution. Western hemlock

A growing room similar to the culture growing room is one option for establishing tissue cultured plantlets (from either Stage II or Stage III) into soil. Light and heat are easily controlled. Humidity is maintained by planting in covered containers. An 11" × 21" seedling tray with a plastic cover (Humidome) is a standard unit, but smaller containers confine disease problems to fewer plants. A transparent 4" × 4" plastic berry cup can hold up to 50 plantlets. A second cup inverted over the top makes a miniature greenhouse. The top and bottom cups are taped together and need not be separated for observation or watering (Fig. 11-d). Drainage holes in the inverted cup afford room for a wash bottle stem to be inserted for watering (although they may need to be taped over to start with). They can also be watered from the base by flooding a holding tray.

On a larger scale shaded greenhouses are equipped with tunnels or tents on benches. These enclosures are variously equipped with mist, bottom heat, exhaust fans, cooling pads, or lighting with appropriate timing and sensing devices. A whole greenhouse might be similarly equipped but moisture and humidity requirements become increasingly difficult to adjust the larger the greenhouse. A fog system is an ideal way of maintaining high humidity without oversaturating the plants.

References: #6, #40, #41, #42, #55, #81, #135, #137, #145, #169.

SHIPPING

Various methods of packing tissue cultured plantlets have been used. To ship plantlets in the glass containers in which they have been growing is costly and the potential for breakage is high. Shipping plantlets in autoclavable plastic containers or aluminum food tray involves less cost and risk. But only rarely is it necessary or desirable to ship plantlets in a sterile condition. A satisfactory method that has worked well for caneberries is to remove the rooted plantlets from jar or rooting tray, roll 50 of them in damp paper towels ("jelly-roll"), pack 10 rolls in plastic bags, and place 6–12 plastic bags in a small carton. The small cartons are packed in shipping cartons lined with sheets of foam to protect against heat or cold during shipment.

Chapter V: Other Pathways

PROTISTA

While no longer considered members of the Plant Kingdom but members of the Protista Kingdom, algae, fungi and bacteria are commonly cultured in vitro.

Algae

Fresh water and marine algae are valued for food, fertilizer, and secondary products. Their interesting life cycles are routinely studied in biology classes. Their beautiful patterns may be an inspiration to artists but their ecological importance is a nightmare to government officials because they congest waterways, lakes, and streams. They also are responsible for causing toxicity in some shellfish, even deadly, to humans. Many algae are microscopic, thus lending themselves conveniently to liquid culture.

Sterilized, filtered soil water (not all soils work) and Hoagland's solution are common media for fresh water algae. Premixed salts for artificial sea water are available commercially for marine algae.

Fungi

For the most part fungi and bacteria go quietly about their business of decomposing organic matter with few people aware of this contribution to a better environment. On the other hand, fungal diseases are not, and cannot, be ignored. They can even change the course of history as in 1845. It was then that the water mold, *Phytophthora infestans* contributed to the Irish potato famine. The disease this mold causes, called "Late Blight" so affected potato production that a million deaths were attributed to the Irish potato famine and another half million Irish migrated to the United States.

Fungi are noteworthy for both their desirable and undesirable relationships with higher plants. Because many fungi are plant pathogens they are commonly cultured and studied for their life cycles and interactions with plants. They are cultured to test plants for resistance or immunity to them using the whole organism or filtrated extracts. Chemical companies, in particular, are obliged to learn how to prevent fungal diseases and how to cure them once they occur. Horticulturists are also interested in mycorrhizae which are fungi that enter into symbiotic (mutually beneficial) relationships with plant roots. The list is without end.

The ease of growing molds is apparent in tissue culture of plants as soon as contaminants appear. Pathogenic or not they will quickly overgrow a plant culture. However, to deliberately culture fungi there are readily available, inexpensive, premixed media that are often used such as potato-dextrose-yeast agar, or "nutrient" agar, or malt extract agar, to mention a few. There are also agars with inhibitors and/or certain pH levels to aid in isolating particular fungi, and diagnostic media to aid in identification. A few oat flakes on filter paper are all that is necessary for growing some slime molds, amebae-like Myxomycetes with fungus-like fruiting bodies.

Mushrooms are a popular food item. While tramping through the woods in search of the savory mushroom may be an appealing alternative to buying mushrooms at the store, there is yet another way of satisfying such tastes, that of growing them yourself.

The procedures are no more difficult than for plants but culturing mushrooms is complicated by interaction with other fungi, both good and bad. An excellent guide to this fascinating field has been written by Paul Stamets and J. S. Chilton, *The Mushroom Cultivator, A Practical Guide to Growing Mushrooms at Home,* (Agarikon Press, Olympia, Washington)

Bacteria.

Some of what has been said about fungi applies to bacteria. Techniques for culturing bacteria are broadly similar to those for culturing molds and yeasts. Bacteria also interact with micropropagation of "higher plants" when they (a) are contaminants in culture, (b) cause plant diseases (it is counter productive to multiply diseased plants), (c) are used in screening for disease resistant plants, (d) are participants in the induction of secondary products, and (e) are involved in genetic engineering.

ADVANCED TECHNIQUES

Micropropagation as described throughout this book is basic and elementary. The requirements for such basic tissue culture are relatively unsophisticated compared with the know-how required to do cell culture, protoplast isolation and fusion, and somatic hybridization. As these advanced techniques come out of research and into practical application the reader will need to be aware of them, understand their usefulness, and have some concept of the procedures involved.

It must be emphasized that these procedures, in terms of real plants waving in the wind, are not easily or often realized even by the most astute students or skilled laboratory scientists. The media formulas and approach are different from shoot tip culture. Careful sterile technique and chemical cleanliness are extremely rigid, considerably more demanding than for simple techniques. Moreover the goals are usually different. Traditionally the objectives of cell culture technology have been the study of cell nutrition and metabolism, disease and pesticide resistance, the production of secondary products, somatic hybrids, haploid culture, protoplast isolation and fusion, genetic engineering, mutagenesis, and other phenomena.

Cell suspension cultures, for example, are a convenient means of studying cell utilization of, and response to, different nitrogen and carbon sources. The data gathered from such studies are important not only for cell cultures but the information also applies to plants not in culture. Cell cultures are used for screening cell resistance to introduced pathogens, herbicides, fungicides, salinity, and other factors. The cells that do not succumb can be grown out to whole plants possessing corresponding resistance.

Cell and callus cultures are used as a source of secondary products. Secondary products are defined here as secondary metabolites: products of metabolism that are available by extraction from plants in culture. The products are usually of a medicinal nature but they also comprise certain food additives and industrial compounds. Some examples are steroids, terpenoids, alkaloids, pigments, flavorings, and hydrocarbons.

One of the more successful secondary products is shikonin from cell suspensions of *Lithospermum erythrorhizon.* Other extracts include biotin from *Lavendula vera, Nicotiana tabaccum,* and *Glycine max;* nicotine from tobacco callus; saponin from *Panex ginseng;* anthocyanin from grape; steroid from yucca; berberine from *Argemone mexicana, Coptis japonica, Phellodendron amurense,* and *Nandina;* L-DOPA from several legumes; alkaloids from *Catharanthus roseus;* and caffeine from *Coffea arabica. The Merck Index* is the ideal reference book to learn about these substances.

Two steps are often used in the production of secondary products. The first step is culture multiplication, then product manufacture by the culture is induced by a second set of parameters such as change of medium, or application of antagonists.

The listed objectives, however, do not rule out the goal of plant multiplication itself by way of cell and callus culture. Cell culture is a productive and efficient method of multiplication, barring mutations that occur. The cells and/or embryos are ultimately plated out (spread on agar medium) and grown-out with or without a callus stage. Some cells will give rise to somatic embryos; in other cases plantlets are only obtained by way of organogenesis, i.e., adventitious shoots from callus. (See summary flowchart below.)

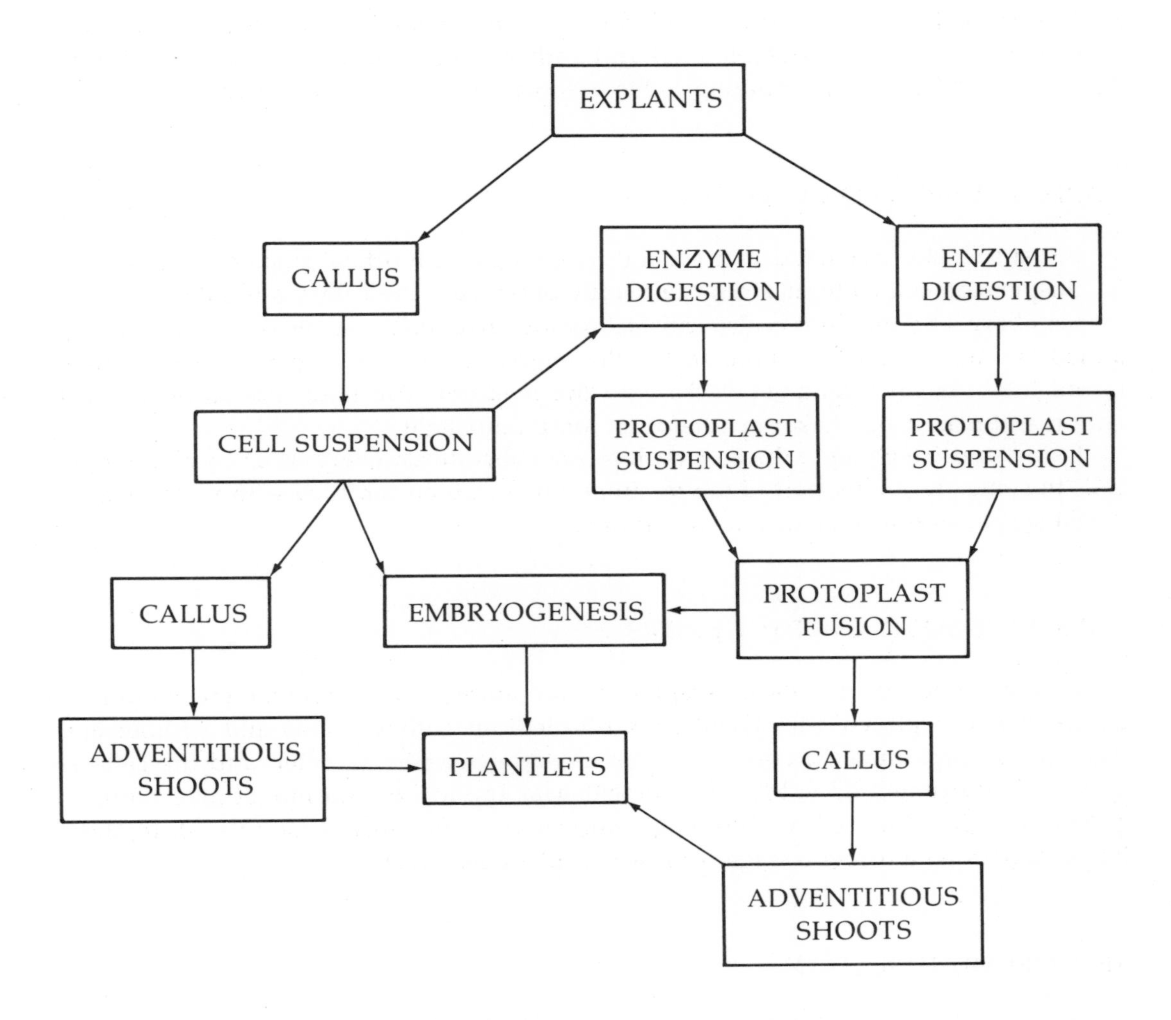

Figure 15. Summary, cell culture flow chart

CARROT CALLUS CULTURE

Historically, carrot and tobacco callus cultures have been the classic models studied. Callus tissue culture was standard procedure for the regeneration of whole plants from culture until the merits of shoot tip and adventitious bud culture were realized.

Because of its historical value and ease of callus inducement carrot is used here for example. The procedures outlined here are not intended as instruction but only for information and example. Please see the references specified for greater detail and excellent guidance.

The first step is cleaning viable seeds in alcohol and bleach, followed by sterile water rinses. They are placed on ½ MS in test tubes or petri dishes and allowed to germinate. After germination 2–4 cm sections of hypocotyl (stem below the first leaves) are placed on Gamborg 5 (or other suitable medium) with 2,4-D and agar. They are placed in the dark at 25°–28°C and the resulting callus is subcultured every four weeks.

CARROT SUSPENSION CELL CULTURE

For suspension cell cultures a rotator or open platorm orbital shaker is required.

Suspension cell cultures are more readily obtained if the callus is friable.

Gamborg's B5 medium is also satisfactory for carrot cell suspension culture. 2,4-D is added and agar is omitted. Four to 6 callus pieces, each about 1 cm square, are transferred into 125 ml Erlenmyer flasks with the medium. The flasks are stoppered with cotton and placed on shaker or rotator in continuous light.

Using sterile technique the suspensions are subcultured every week by allowing the cells to settle, decanting part of the medium which is then replaced with fresh medium. A cell suspension usually develops within 6 weeks.

CARROT SOMATIC EMBRYOGENESIS

To condition carrot cells in suspension for embryo development the medium is decanted and replaced with Gamborg's B5 medium without 2,4-D and without agar. Agitation is continued. In two weeks the medium is again decanted and the cells are spread on ½ strength B5 or MS medium with agar and without hormones in test tubes or petri dishes. Dividing cells undergoing embryogenesis first form a globular stage, then a heart stage, then a torpedo stage from which plantlets develop.

PROTOPLAST ISOLATION

Protoplasts are plant cells which lack their normal, rigid, cellulose walls but still retain their plasma membranes. They are a valuable research tool and hold promise for breeders. Because they are without walls many substances are readily taken up through the fragile membranes including viruses, nuclei, chloroplasts, DNA, protein, other macromolecules, and other protoplasts.

Most growers will not want to attempt protoplast culture but it is important to understand the principles and procedures to appreciate their potential.

Starting material may be any rapidly growing tissue cultured callus or shoots, cell culture, or leaf mesophyl from young, fully expanded leaf.

Following are broad statements of protocol for protoplast isolation assuming in vitro starting material:

1. 1–2 day dark pretreatment.
2. Cells are centrifuged in Gamborg B5 solution
3. Incubated with filter-sterilized enzyme solution (cellulase and Macerozyme) 1–12 hours.
4. Filtered through 60 μ screen.
5. Centrifuged.
6. Observed for protoplast release.
7. Added to appropriate culture medium.

If the starting material is a leaf it is cleaned in bleach, and rinsed. The epidermis is peeled off and the enzyme solution applied, followed by steps 4–7.

PROTOPLAST FUSION

By learning to remove cell walls one barrier to intergeneric crosses was overcome. The second barrier was the lack of fusion between protoplasts from different species or genera.

It was a big step forward in plant tissue culture when, in 1974, Kao and Michayluk discovered that polyethylene glycol could induce fusion of protoplasts from different genera or species. The extreme results of this discovery are the fusion (or uptake) of protoplasts of higher plants with protoplasts of lower microorganisms; in fact, plant-animal fusions have even been observed (none have survived).

Conventionally, in protoplast fusion studies the protoplasts from a cell suspension culture form callus, which is not green but whitish, and a cell suspension culture from leaf mesophyll cells, which is green, are used. Such a mixture facilitates observation of fusion; i.e., due to the color difference which can be observed when similar or dissimilar cells are fusing.

HAPLOID CULTURE

Haploid cultures are derived from pollen usually by way of anther culture. Haploid plants are of value to plant breeders and geneticists because there is only a single set of chromosomes to contend with. If chromosomes spontaneously double, or are induced to double with colchicine, the double haploid plants that grow are potentially desirable for breeding purposes because the two sets of chromosomes are identical.

The ease of spontaneous doubling in some species can be considered an asset or a drawback depending on one's objectives. Some species are known to possess a degree of non-haploid pollen which is heterozygous (double but without identical chromosome sets), therefore useless. If whole anthers are cultured, vegetative cells (also heterozygous) are present. Haploid plants are often distinguishable by their smaller size. At an earlier stage counting the chromosomes will determine if the cells are haploid, if not confused by chimeral differences in chromosome number which occur in callus culture.

Anther culture procedures vary significantly from one species to another and from one lab to another. The failure rate is high. The time for obtaining anthers is critical because the stage of pollen development is crucial to success. For example, a rule-of-thumb for tobacco is that the anthers are ready when flower bud sepals and petals are the same length. Buds are pretreated at 8°C overnight, cleaned in alcohol and bleach, and the anthers removed. Placed on appropriate medium under fluorescent light (18 hour photoperiod) plantlets develop in 6 weeks.

Recommended reading: *Handbook of plant cell culture.* D. A. Evans, W. R. Sharp, P. V. Ammirato, and Y. Yamada, editors; Macmillan, 1983.

Cell Culture and Somatic Cell Genetics in Plants. I. K. Vasil, editor. Academic Press, Inc., New York, 1984.

Tissue Culture as a Plant Production System for Horticultural Crops. R. H. Zimmerman, R. J. Griesbach, F. A. Hammerschlag, R. H. Lawson, editors, USDA. Martinus Nijhoff Pub., 1986.

References: #3, #4, #39, #48, #114, #127, #140, #153, #158

Chapter VI: Tissue Culture Business

COST PROJECTION—LAYING THE GROUNDWORK

In order to adequately judge the financial success of a commercial operation it is necessary to project the financial picture in advance, to record it in progress, and to periodically evaluate it in retrospect. This book is devoted to tissue culture and not to business management, but there are certain questions to consider in order to determine the utility of a tissue culture lab.

Because it is not possible to foresee how large an operation may become, expansion potential should be built into the original planning; otherwise a piecemeal, expensive, disorganized assortment of additions will result. A local contractor can help determine facility cost, but shopping for oneself with a detailed plan in hand can result in significant savings. Equipment can be changed more readily than a building. "Scientific" equipment is fine but oftentimes household equivalents can be adapted, for example, a kitchen dishwasher instead of a lab glass washer, or a pressure cooker instead of an autoclave. Alternatives should be studied and balanced against the size and nature of the operation being considered. In other words, do not build a lab for 150,000 plants annual production and buy equipment for a million plants, or build a lab for a million plants and try to get by with household equipment. Many other considerations and questions come to mind.

One must be familiar with the appropriate markets. What is the market for the anticipated product of the lab? Who will grow it on? Will there be more profit in selling plantlets, liners, or gallons? How much will each of these end products cost and return in profit? Is it really cheaper to tissue culture the varieties in question than to maintain stock plants and grow from cuttings? What is the time factor for establishing a variety in culture, i.e., how long will it take to reach Stage II, the multiplication stage? Boston ferns may take a month, strawberries two months, and rhododendrons six months. What is the multiplication rate and the rooting period? Is it cheaper to build a lab and produce the plantlets than to have them custom tissue cultured in an existing lab? Answers to these questions will help to answer the question: will there be a profit?

Cost Studies

Cost studies have been published on lilies and broccoli (Table 4). In the lily study one supervisor, three technicians, and a greenhouse technician were involved in producing 33000 Red Carpet lily plantlets. Note that three times the number of lily plants were produced as broccoli plants within the same time span. Indirect costs were calculated on a weekly production basis instead of cost per square foot because space was considered a relatively minor cost. For this reason equipment, laboratory, and overhead cost allocations are not the same for both crops even though the time span is the same. The cost of lily plants is less than the cost of broccoli plants because of the greater numbers and ease of handling of the lilies compared with the broccoli plants in the same length of time. The lily calculations conclude after planting into Jiffy 7's, at which time they presumably were sold.

The broccoli operation involved one supervisor, four technicians, and a greenhouse technician. Premixed MS salts and additional constituents (added in the lab) were used in both the lily and broccoli examples. One technician averaged 27.8 hours per week for making media and cleanup. In the greenhouse stage 275 broccoli plantlets were transplanted to Todd Planter Flats per hour. If Stage III had been omitted and the plantlets rooted in the greenhouse, $0.03 per plant could have been saved.

Table 4. A summary of two cost studies by Anderson and Meagher[#9, #10]

Operational data	*Broccoli*		*Lilies*	
Explants	147 flower buds		43 scales	
Transfer cycles	5		5	
Total weeks	24		24	
Rooted plants	11000		33000	
Transfer hours	132		66	
Tissue Culture Expense				
Wages	$ 475		$ 306	
Media preparation and cleanup	$ 100		$ 72	
Supervision-Administration	$ 394		$ 394	
Total wages and salaries		$969		$772
Media cost	$ 69		$ 35	
Culture containers amortized	$ 17		$ 8	
Equipment, $10000 amortized 5 years	$ 47		$ 40	
Lab facility, $50000 amortized 20 years	$ 103		$ 48	
Overhead	$ 66		$ 66	
Culture cost	$1271		$ 969	
Culture cost per plant	$0.116		$0.029	

Greenhouse finishing cost		*Transplant to Jiffy 7's*	
Wages	$ 144	Wages	$ 108
Materials	$ 31	Jiffy 7's	$ 265
Bench charge	$ 241		
Total	$ 416	Total	$ 373
Greenhouse cost per plant	$0.038	Transplant cost per plant	$0.011
Total cost per plant	$0.154	Total cost per plant	$0.041

The two cost studies cover salient points that bear attention before a lab is built, and periodically, during operation. Although the studies were done in 1977–1978 they are still valuable and valid when a current dollar adjustment is made.

Crop planning

As part of overall planning, a lab director also correlates space and transfer room activities with production demand. For crops with a regular, short-term turnover, Boston ferns, for example, planning is relatively easy. For crops that build up to 30 to 50 thousand for spring delivery, space management becomes complicated. A good plan starts with the end delivery date and number ordered. Working backward toward the explants, target numbers are established and the space requirements noted for given points in time.

CROP PLAN

Hood strawberries		Order #852		5000	Due May 1, 1983	
Date	*Number*	*GH space*	*Jars*	*Tubes*	*Shelf, sq. ft.*	*Transfer labor hours*
Deliver May 1	5000					
To GH by Mar 1	5200	190 sq. ft.				75 GH
To rooting by Jan 1	5500		220		16	10
Multiply Dec 15			80		9.2	3
Multiply Dec 1			25		1.8	1
Multiply Nov 15				128	2.5	1
Multiply Oct 15				42	1	.5
Explants by Aug 15				25	1	2

To this sample plan could be added any other factors deemed important to the planner such as media needed, equipment required, and costs.

After a crop plan has been established for each variety, an overall space schedule is made to assure adequate room and product flow. Usually, as numbers increase cultures are relocated. When these moves are planned ahead there is a minimum of confusion and wasted motion.

SAMPLE SPACE SCHEDULE

Rack	*Shelf*	*Aug.*	*Sept.*	*3 Oct.*	*Nov.*	*Dec.*	*Jan.*
I	1					STO	
	2				HTG, STO	HTG	
	3	WOL, WSH STO, LLM CEN, HGT	WOL, WSH STO, LLM CEN, HGT	WOL, WSH STO	WOL, WSH	WSH	
	4			LLM, CEN HGT	LLM, CEN	CEN	
	5					LLM	
II	1					WOL	

In addition to the space schedule a transfer schedule should be made listing plants and the week they must be transferred. Scheduled transfers are indicated by an X which is circled after the transferring is finished.

SAMPLE TRANSFER SCHEDULE

	Sept.				Oct.				Nov.				Dec.				Jan.
WOL	x			x				x				x			x		
WSH		x			x			x			x			x			
STO		x			x		x			x			x		x		
LLM		x			x			x			x			x			
CEN			x				x				x				x		
HGT	x			x			x			x			x		x		

When you have about enough plantlets to fill your order a simple system will tell you how many to put into rooting and how many into multiplying in the last two transfer cycles. For example: you need 25000 plants. At 30 plants per rooting jar you will need 833 rooting jars for 25000 plants. You have on hand 200 multiplying jars. They have been multiplying at the rate of two to one. You judge that each multiplying jar will make three rooting jars (remember that the clumps will be separated as they are transferred to rooting jars). Thus your 200 jars will make 400 multiplying jars or 600 rooting jars. This is too many multipliers (1200 rooting jars next transfer), and not enough rooters at this transfer so determine how many of each should be turned out this transfer in order to have 833 rooting jars after the next transfer.

Make a chart with jars to go to multiplying on one axis and jars to go to rooting on the other. The sum where the two axes meet is 200 in this example. The chart could be run from zero to 200 on both axes but a likely range is illustrated.

JARS TO ROOTING	JARS TO MULTIPLYING 60	70	80	90
140	120 mult. × 3 = 360 R 420 R 780 R			
130		140 mult. × 3 = 420 R 390 R 810 R		
120			160 mult. × 3 = 480 R 360 R 840 R	
110				180 mult. × 3 = 540 R 330 R 870 R

From the chart it is easily determined that if 80 of the 200 jars are transferred to 160 multiplying jars their next transfer will produce 480 rooting jars. The remaining 120 of the 200 multiplying jars on hand will make 360 rooting jars this transfer. With this combination (80 to multiplying and 120 to rooting) 840 rooting jars will have been produced by the end of the next transfer.

As you plan your lab you will find it very enlightening to make up a set of schedules such as those presented here. At first it may appear to be a futile exercise, but before you have finished you will realize that you can visualize the total operation much more clearly. You will be better prepared to anticipate challenges and evaluate whether or not tissue culture is the business you wish to pursue.

Cost of contamination

One of the greater challenges in tissue culture is contamination. It is an important economic factor and it can play havoc with a good plan. A culture run starting with 1000 explants which is expected to produce 40000 plants will only produce 37600 if 2% become contaminated in each stage. At a price of $.35 per plant, the lost profit will amount to $840. This example further emphasizes the importance of planning with an eye to sanitation. Money spent for air filtering, sterile transfer, and containers that prevent contamination is money well spent.

Refrigerated storage

Refrigerated storage of cultures is a tool that can be incorporated either deliberately in original plans or as a recourse to ease workload when there is too much to transfer at one time. If cultures become aggressive and are ready for transfer ahead of schedule it is possible to refrigerate some species until it is time to transfer them. Strawberries, rhododendrons, carnations, and chrysanthemums are some of the genera known to tolerate several weeks or months of storage at 34° to 36°F, usually without any light.

Energy efficiency

Although natural lighting has been used by some tissue culture labs in the growing room, it is not recommended. Although it is economical, natural light fluctuates radically, therefore, it is difficult to manage. Fluorescent lighting, on the other hand, is constant, relatively inexpensive, and provides some or all of the heat required in the growing room. It is more economical to schedule the dark cycle during the day, when heating requirements are low. If the light cycle is scheduled during the night, heat from the lights can be circulated through the growing room and lab, minimizing the use of conventional heat sources. Even when ballasts from fluorescent fixtures are relocated outside the growing room, to prevent overheating and reduce cooling requirements, they still provide a heat source which can be tapped as required by the air circulation system.

Common energy saving practices such as good insulation, tight doors and windows, and turning off lights and appliances should obviously be observed.

COSTS AND LABOR

Because labor represents the greatest expense in the operation of a lab, management must watch constantly for ways to maximize labor efficiency. The cost of premixed, or partially premixed media should be studied and weighed against the cost of labor to mix it; however, the desirability, or perhaps the necessity, of having the flexibility of varying the ingredients in the lab should not be underrated.

Labor saving devices are usually a good investment. Media dispensers, automatic labelers, and instant read-out balances are examples of equipment which can help maximize labor output.

Travel time represents dollars whether from one room to another or the distance and frequency the hand travels in operating equipment or making a transfer. Pass-through windows save time by eliminating the need to carry or cart cultures and materials through doorways. Similarly, careful organization of equipment around the transfer technician contributes to the best use of transfer time.

The size and shape of culture containers is of great economic concern. Tubes are valuable for starting cultures and restricting contaminants. However, larger containers are far more economical in terms of transfer time. The adept technician can transfer over 128 tubes an hour when dealing with good material and a 3 to 6x multiplication. However, transferring into pint jars at a rate of 20 jars per hour, each with 16 propagules (pokes or pieces), is equivalent to 320 tubes per hour. John Song (Magenta Corp.), has arrived at a two cents per plant sold lower labor cost by transferring into a 3″ × 3″ plastic container with a 3″ × 3″ cover (autoclavable) as against transferring into a Mason jar used sideways and closed with foil circles. The plastic container, he claims, is easier and cheaper to use. The plastic container costs nearly three times as much as the jar, but once labor savings have offset the higher container cost, further labor savings theoretically become profit. Admittedly, this presents a difficult choice for the person starting a lab and trying to hold down the initial investment.

During the periods when transferring activity is reduced, due to small lab inventories, the opportunity to rotate personnel from nursery to lab is ideal for the established grower. Lab crops targeted for spring outplant make winter the busiest time of the year in the lab, usually a slow time in most nurseries. The advantage to retaining year-round employees is evident.

While providing the potential for physical expansion of the lab it is well to consider growing personnel needs as well. The preliminary cost projection should stipulate the type of employee a tissue culture lab unit must hire. Depending upon its size and projected output, the range of employees for nursery tissue culture ranges from laborer to Ph.D. Personnel must already have, or acquire, some knowledge of tissue culture, sterile technique, chemistry and plants.

Familiarity with tissue culture information on the cultivars propagated can be gained by reading the periodical literature, from work at college or government labs, or from books such as this. Sterile technique is taught at some colleges but is described in a variety of books, and can be quickly self-taught with practice and imagination. People with a solid high school chemistry background can follow a formula for mixing media, or teach someone else with no training to do so. Academic training can help bring these factors together if other equally important qualifications such as maturity, common sense, organization ability, and business or production experience are possessed by a particular individual.

If no one within an existing nursery organization is qualified or available for training to direct the lab, advertising in trade journals can frequently turn up the right person for the job. College placement offices can assist in identifying technicians. Often, competent personnel can be located at meetings of the International Plant Propagators' Society, The International Association for Plant Tissue Culture, or the Tissue Culture Association (see Appendix, page 142).

Salaries for relatively untrained help ranged from $800 to $1200 per month in 1987. Salaries for academically trained or experienced tissue culturists ranged from $1200 to $2500 per month and up.

Regardless of background, the lab employee must realize the difference between research and production. The inquisitive will want to experiment, the overly exacting will want to pick at transfer material or dawdle with labeling or records. There is no faster way to lose profit. Problem material will, of course, require special treatment and slow up production; it should not be left to the amateur. Efficient mainstream production is at the heart of nursery tissue culture.

COSTS AND SPACE

Space is not as tangible a cost as wages but it is a real expense. With a tissue culture operation space savings are realized in several areas of nursery operation. First, less space is required to carry stock plants. Secondly, a tissue culture lab and growth room require significantly less space than do the facilities for starting seedlings, cuttings, or grafts. As a consequence fewer dollars need be invested in land and buildings. And lastly, significant savings can be realized from the use of limited growing space through scheduling of propagating/growing activities to which culture lends itself so handily. In short, the grower need allocate fewer dollars to space and realize a higher return on the dollars invested in land and buildings.

RECORDS

As cultures are started, multiplied through several transfer cycles, and eventually moved out of the lab into other nursery space, it is essential to keep the right records. The careful records kept for a research project are not usually economically justifiable for production purposes. At one end of the record keeping spectrum is the identification of each explant with the particular source plant and maintenance of the identity throughout the history of the clone. Where a number of clones of the same cultivar are involved, the advantage of such a meticulous record helps establish the easy elimination of a clone should a problem of cultivar identity or an anomaly develop in either the clone or the source plant. As the number of source plants and explants increases, however, the complexity of record keeping increases. The grower must weigh the insurance value of extensive record keeping against the cost of producing a faulty product. The object of record keeping in production tissue culture is to provide sufficient cross checks for peace of mind without creating costly busy-work. One of several records which should be kept is a simple log, date book, or diary.

LOG

Aug. 26	KS Bentons ahead of schedule by 20%.
	Mixed media for raspberry hormone grid.
Aug. 29	Started test on sterile paper towels.
	Replaced prefilters in transfer hood.
Aug. 30	Painted new shelves—3 hours.
Aug. 31	Calibrated pH meter and thoroughly cleaned electrodes.
	Dr. Z. visited.
Sept. 2	Ordered glassware from Am. Sci.
Sept. 3	Called DJ his plants are ready.

A record of transfer information is essential. One simple filing system depends on a 3" × 5" card filed for each cultivar. In addition to the name and code for the cultivar, it carries the customer's name, date started, transfer dates, culture conditions, number transferred, medium used, tests performed and results. An example follows:

C202
D. Johnson
Rhododendron

1981

June 18– Started 16 explants. 10 min 1/100 cx plus 2 drops Twn. DH_2O rinse, 10 sec alc., rinse, 15 min 1/10 bleach with 2 drops Twn, 3 rinses. To liq. RH.

June 22– 3 moldy, redid in 50% bleach for 5 min.

June 30– Discarded 4. Tsfr 12 to 12. Fair. 6/22 ok.

Aug. 15– Tsfr 12 to 15. Rh agar. 3 good shoots.

Another valuable record is a chemical roster noting when a new container of chemical is opened, when and where it was purchased, the cost, and where it is stored (drawer, shelf, refrigerator, or freezer):

Chemical	Date Opened	Date Purchased	Vendor	Cost	Storage Location

Finally, a carefully set up and maintained book of general financial accounts is essential. You can lose money quickly on a tissue culture operation. It can be a year or more before any money comes in, but if and when it comes in is important. Patience is the password, but at some point you must answer the question, "Are we making any money?" A good set of financial records, meticulously maintained will answer this question and is the key to successful financial management.

With the computer age upon us, do not overlook the possibility of computerizing tissue culture records. The system of schedules, inventory, and records lends itself very neatly to computer programming. Its cost effectiveness for each individual nursery bears investigating.

Using a commercial program entitled "pfs FILE" a satisfactory transfer tracking record has been worked out for an Apple IIe computer. The following form is completed by the supervisor or the person transferring:

Date next transfer:
Date today:
Code:
Number in:
Number out:

Number discards:
Reason:
Condition—comments:
Time:
Technician:

With this data on a floppy disc it is easy to call up the codes of those cultures due for transfer on a specific day or week, history of a clone, or any other information contained on the file.

MAKING THE PROJECTION

This book is about tissue culture, not nursery management. The foregoing thoughts have simply been furnished to help you with some sense of the unique characteristics of tissue culture. These ideas, coupled with professional advice on business planning and projections, should provide a firm foundation, a base from which to answer some of the questions posed earlier and to decide whether or not to start a lab.

References: #6, #7, #8, #9, #56, #117, #122, #139, #144, #151, #153, #162.

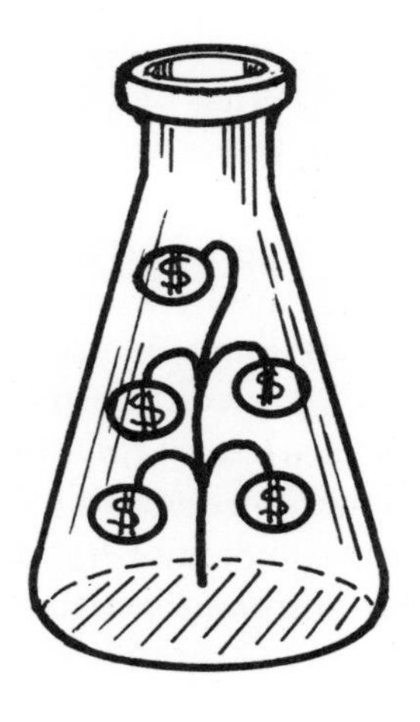

Chapter VII: Tissue Culture Potential

Tissue culture is a field of many facets. It varies from the curious gardener working in a modest home kitchen to the renowned scientist working in an elaborate laboratory. It reaches from the orchid hobbyist who has learned to multiply a few favorites to a million dollar industry producing pyrethrums. Specialists are learning to vegetatively hybridize plants from entirely different families. Genetic engineers are discovering how to identify and manipulate genes, how to remove individual characteristics from one plant and introduce them into another.

The spectacular achievements behind closed doors of elite laboratories do not prevent the curious minded from indulging in a fascinating hobby. The large undiscovered world of tissue culture is waiting for exploration. Hobbyists, who are not bound by the constraints of production, are free to pursue any avenue their aptitudes, time, and budgets will allow. Each will view the field with a unique background and insight whether it be from academic or practical experience. Some will have studied chemistry, horticulture, or microbiology; others will have studied the ways of plants in field, greenhouse, or garden; all will have two traits in common, wonder and curiosity.

The simplest tissue culture hobby is the multiplication of easy, fast growing subjects such as Boston ferns, African violets, or Rex begonias. Next, in order of complexity, are carnations, strawberries, or *Syngonium.* Much is still unknown about the culture of even the most commonly tissue cultured plants. Culture media are always subject to change and improvement. Techniques need to be carefully described in detail, they need to be improved for commercial labs, and the information needs to be made freely available to potential users. Moreover, there are the countless plants that have never been tissue cultured, many of them endangered species. What better legacy than to save some of these precious flora by means of tissue culture? These, and many other areas of effort are open to the amateur.

The more studious and affluent plant hybridizer might be attracted to hybridizing by way of protoplast fusion. The process involves removing cell walls with enzymes followed by treating the naked protoplasts with polyethylene glycol, which induces fusion. Relatively few plants have been grown to maturity, but the potential is enormous. The procedure requires considerably more lab know-how and equipment than simple micropropagation. It need not, however, be confined to the laboratories of academia.

The chemist who is looking for a tissue culture hobby may be challenged to explore the field of plant by-products in tissue culture. Flavorings, medicinals, oils, and dyes are a few of the products that are being sought from biosynthesis in plant tissue culture.

Gardening is one of the most common hobbies. Conventional gardening is limited by the seasons of the year, but tissue culture knows no season. Gardeners who propagate by tissue culture will delight in year around micropropagation. If successful, they may find they have more plants than they anticipated. Excess plants are shared with friends or offered for sale, and many hobbies have turned into businesses.

People who are physically unable to garden may find pleasure in watching and caring for plants in culture. There is little physical exertion required in the comfortable climate of tissue culture operations.

The hobbyist or amateur gardener need not feel limited in pursuing tissue culture for want of a transfer hood and lab required by a commercial operation. They can very well use make-do equipment. Small scale tissue culture is often carried out without benefit of lab or special equipment. It can be done by almost anyone in almost any house. Transferring on a desk or table in a clean room free from dust and drafts is feasible. Transferring in a home-built chamber with a glass or plexiglass front (Figure 8-3) with just room enough for gloved hands to enter, is a reasonable method for the hobbyist.

Premixed media, a pressure cooker, forceps or tweezers, a paring knife, a few test tubes with caps, or Mason jars, household bleach solution, a lighted shelf, and a lot of determination will bring exciting discoveries to the amateur.

In contrast to the hobbyist, the commercial grower is compelled by finances to make tissue culture a profitable enterprise. Growers with limited resources who must make a living from a small operation are finding that vast numbers of plants can be propagated by tissue culture with minimal space and outlay of capital. Oftentimes they tissue culture one or two cultivars consistent with their operation and build a reputation for these cultured specialties. A few of the plants that have made reputations for their growers are carnations, ferns, iris, fruit tree rootstock, and rhododendrons. In most cases these growers have used tissue culture propagation for in-house growing-on, but sometimes other growers welcome the opportunity to buy tissue cultured, rooted plantlets, especially if they are hard-to-start or hard-to-find plants. The latter situation conserves space and allows a quicker return on tissue culture investment. However, in-house growing-on is important to the nursery's normal market and precludes the competition which may be generated by selling plantlets to other nurseries.

In lieu of building their own labs, some small scale growers with growing-on capability decide to buy rooted plantlets from established commercial tissue culture laboratories. There is room in the industry for the grower who will simply take plantlets directly from a commercial lab and harden them off to where they or other growers can finish them to sell.

Growers, plant brokers, salespersons, greenhouse suppliers, or others interested in marketing, have a new product line in tissue cultured plants. The new markets are (1) plants still in culture, (2) plants directly out of culture, rooted or unrooted, and (3) plants hardened off to greenhouse conditions. Tissue cultured plants are always true to type, more vigorous, more disease resistant, and disease free. When the end product costs less than conventionally propagated material, tissue cultured plants sell themselves.

"Pathogen free" plants in culture open the door to more freedom of exchange of plants between states and with foreign countries as well. Plant tissue cultures have gained acceptance in world trade as the danger of introducing disease is virtually eliminated. Foreign exchange will increase as new hybrids are developed asexually from protoplast fusion, and other feats of genetic engineering find practical application. Free exchange of tissue cultures will have a significant impact on world food problems by bringing more improved cultivars more rapidly to growers everywhere than in the past.

The application of tissue culture to farm crops has scarcely begun. A significant beginning has been made with tissue cultured fruit trees. Fruit trees are difficult to propagate from cuttings. To maintain clonal characteristics, desirable scions are grafted onto special rootstocks usually propagated by layering or cuttings. Rootstocks are cloned to maintain the known influence of particular rootstocks upon scion growth such as hardiness, disease resistance, or dwarfing. Commercial tissue culture labs are currently culturing rootstocks very successfully. If tissue cultured rootstocks prove to be economical, they will probably replace layered rootstocks. In fact, fruit trees are being tissue cultured and grown on their own roots. If successful, grafting will become a cumbersome process of the past.

Secondary Products

Commercial application of tissue culture is also found in secondary products. A fascinating example is the gigantic pyrethrin industry in Ecuador. Each month 62,000 tissue cultured pyrethrum plantlets (*Chrysanthemum cinerariaefolium*) come out of the company lab for further propagation and growing-on. The 11 million plants that are planted annually in the high Andes provide a living for hundreds of villagers who pick the flowers for extraction of insecticidal pyrethrins.

Although few secondary products are actually being produced in vitro commercially it is not a question of whether they will be produced by cell culture but when. Hundreds of products being extracted from whole plants are presently and with considerable effort being sought by way of cell and callus culture. The hope is that numerous natural products, now in short supply and of great value to industry, may be produced not only in great abundance but also more efficiently. At this writing a number of cell lines are demonstrating yields equal to or higher than those derived from the original plants.

Not surprisingly media nutrients and hormones have a bearing on production of these secondary products, i.e., alkaloids, pigments, dyes, medicinals, etc. Often cell media, which have been developed for maximum culture growth, are not optimum for production of secondary products. To overcome this obstacle cell populations are first multiplied using conventional media, then the formulas changed to induce the chemicals desired.

In this vein one very exciting aspect of today's biotechnology is referred to as "microbial insult." A filtered extract of a fungus, a *Pythium* species, was added to a cell culture of *Bidens pilosa* (a member of the Compositae family). Presence of the filtrate induced the cells to produce secondary products (certain antibiotic phytochemicals) that were not produced when the filtrate was absent. This example illustrates a new approach to the commercial production of secondary products. It invites speculation not only on the incredible potential of cells, but also on the infinite number of microbial and other inducers that must exist and the diversity of biochemical products they might induce.

New Forests

The demand for timber, pulp, paper, and wood derivatives can only increase as populations grow, but forests the world around are being overutilized and consequently destroyed. Reforestation has not kept pace with the steadily growing demand for wood products. With the urgent worldwide need for reforestation seedlings (preferably from superior trees) it would seem that tissue culture might provide an obvious answer. Millions of dollars have been spent on research to solve the many riddles of propagating superior forest trees via tissue culture, particularly coniferous species.

Micropropagation of progeny stock (stock of known parentage) obtained from seed orchards is, at last, at hand because embryos, cotyledons, and juvenile buds, needles, and fascicles have responded to culturing far more readily than explants from mature trees. Sommer, Brown, and Kormanik were first to report the successful propagation of pine plantlets obtained from pine embryos multiplied in vitro (1975). Even with these advances, however, the state of the art is still plagued by slow growth, aberrations, poor rooting, vitrification, plagiotropic growth, genetic instability, and "half sibs" (where only one parent is known). Both the difficulty in cleaning older material and lack of response by mature tissues have repeatedly blocked the way to micropropagation of select conifer trees.

Growers are well acquainted with the problems of lack of juvenility in rooting cuttings. In many cases the methods applied to retain juvenility or induce rejuvenation for purposes of cutting wood can also apply to obtaining juvenile material for explants for tissue culture. Hedging, or shearing, has long been practiced to retain trees of particular species in a juvenile state in order to provide cuttings which will root.

Alternatively, rejuvenation of mature stock of some species is sometimes accomplished through sequential cuttings or grafts. If a mature shoot can be rooted, the wood produced supplies yet another cutting to be rooted. This process is repeated until juvenility is restored and an explant responds in culture. Another route is to graft a mature shoot on young root stock; the active scion in turn supplies a scion for another graft and the process is repeated, each successive graft hopefully incurring increased juvenility. Still another method, and one with interesting implications, is to repeatedly apply a cytokinin to appropriate nodal areas. This technique has been known to produce witches'-brooms in which the shoots display some degree of juvenility and consequently respond as tissue culture explants.

Assuming "full sib," (where both parents are known), superior trees in a seed orchard is it better to multiply these trees by tissue culture for plantlets to increase the orchard and use seed from that orchard for reforestation or is it better to reforest directly with the tissue cultured plantlets? The economics to answer this question have not yet been resolved. (Fig. 16)

One of the newer approaches to tissue culture of mature trees is the induction of somatic embryogenesis from cells of the nucellus or other tissues of the immature seed. In theory if the embryo is removed from a seed the tissues that remain should be of the same genetic makeup as the mother plant in contrast to that of the embryo. Because of the ease of cleaning seeds, and the juvenility of nucellus tissue, this is certainly a promising approach to micropropagating mature trees. Depending upon the extent of development of the seed the main problem is in ascertaining that the cells obtained are indeed of an identical genetic makeup to those of the mother plant. The subsequent cell culture with ensuing somatic embryogenesis is classic procedure. How much variability will ensue from this technique awaits maturation of these trees.

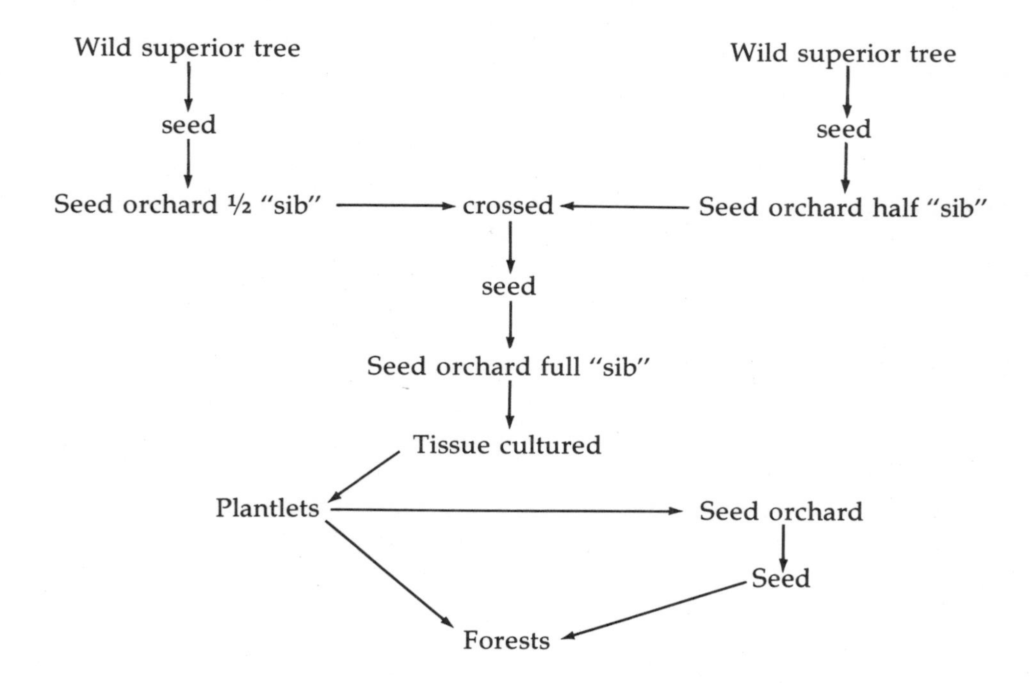

Figure 16. Forest conifer tactics; which is the better route?

With respect to hardwood forest trees the first organogenesis in woody tissue culture was observed by Gautheret in 1940 when he induced buds from cambial tissues of elm (*Ulmus campestris*). The next hardwood milestone was not reached until 1968 when L. Winton reported the first true plantlets from aspen. To date other micropropagated forest hardwood trees include acacia, birch, sweet gum, Paulownia, Santalum, teak, and Eucalyptus. Most of these have been successfully cultured from mature trees, in addition to more "routine" success from juvenile sources.

Progress in discovering what is the potential for micropropagated forest trees and what the options will be, rests on the curiosity, resources, education, and financial incentives of the individuals and organizations involved. One must be alert to the advantages and disadvantages of clonal forests. Costs of micropropagated transplants have been estimated as up to 3000% over conventional production. Further, the potential gain in quality and production of clonal forests must be weighed against the risks of mutations, disease or pest susceptibility. Some answers are just beginning to surface in the few plantations that presently exist.

Cryopreservation

Historically the world's plant germplasm has been confined to "field genebanks," primarily breeders' collections in plantations, orchards, and gardens; and to seed genebanks for storing seeds which lend themselves to drying and maintenance at low temperatures, such as the U.S.D.A. Seed Storage Facility at Ft. Collins, Colorado. The concern among conservationists is that such collections are inadequate.

The encouraging in vitro work of Galzy (1969) and Nag and Street (1973) prompted a report for the International Board for Plant Genetic Resources (IBPGR) on the potential of in vitro storage as a third major means of storing germplasm. Galzy successfully preserved grape shoot tips in vitro at low temperatures. Slow growth continued under refrigeration and returned to normal growth when cultures were returned to room temperature. Nag and Street successfully obtained embryogenesis from carrot cells that had been frozen. The report triggered a world-wide network of cooperation among participating labs doing in vitro cryopreservation (freeze-preservation) and related research.

IBPGR established and operates a germplasm data base. Bona fide enquirers may use the data base at no charge to learn of research, germplasm sources, etc. Particular crops targeted for in vitro cryopreservation include sweet potato, taro, cocoyam, banana, plantain, coconut, sugarcane, cacao, and citrus. IBPGR and cooperating labs are emphasizing disease indexing and eradication, and identification (characterization) of clones in the various collections (germplasm repositories).

Future events

We can also look forward to changes in greenhouse practices due to discoveries in the laboratories. Hydroponics comes close to tissue culture in that plants are nourished by defined nutrient solutions. What can we take from tissue culture and advantageously apply to hydroponics? To what degree can we manipulate plant growth in the absence of sterile technique? The effects of gibberellic acid are well known. Cytokinins, applied to greenhouse plants in a very special manner, produce clusters of mini-cuttings which root in soil. It will take many minds, many years of dedicated work, to answer even some of our questions:

Can it be that someone will develop some frost-resistant tropicals? Will scientists convey the legume capability for nitrogen fixation (with symbiotic bacteria) to grain, to corn, or potatoes? How many plants resistant to how many diseases are in store for us?

How long will it be before we can grow strawberries without strawberry plants, oranges without trees, or manufacture chocolate or coffee without cacao seeds or coffea beans?

Such feats are not for a grower's modest lab. But, we can, each in our own way, be a part of the exciting field of tissue culture. Only science-fiction experts dare predict the future but yesterday's future is here for us today to wonder, to grasp, and to move forward.

References: #3, #21, #46, #50, #65, #96, #114, #125, #150, #151, #153, #168.

SECTION II

Culture Guide to Selected Species

Chapter VIII: Introduction to Culture Guide

There are a number of miscellaneous comments to be made that bear on the following instructions and formulas (recipes). The main point to stress is that there is usually more than one way to make an explant or culture respond. In reading the following the novice may consider them binding and infallible. Actually, quite the opposite is true. They are, at best, approximations of what has worked for someone. Very seldom do two cooks produce identical meals using the same recipes. The same is true of tissue culture and the reason is the same in both cases: there are too many variables for the results to be identical. Used as guidelines, and following the general, basic principles outlined in this book, each is a place to start. The formulas given are derived from the references and other sources of information. The reference numbers correspond to the numerical listing of the Bibliography.

If you don't have all of the unusual additives, leave them out and see what happens. It is amazing how many cultivars will grow on just ½ strength MS plus 1.5% sugar, perhaps not as well as researchers describe growing in more complete media, but well enough to get by. If the optimum temperature given is 23°C the chances are good that the cultures will do just fine in temperatures fluctuating between 20°–26°C. If recommended cleaning calls for 0.1% Tween 20, add 2 drops per 100 ml of cleaning solution and you will be in the ballpark. You will soon learn the variables that work for you.

As the state of the art advances, more plants are being micropropagated without first producing a callus stage. The direct production of multiple shoots is not only more efficient but the plants are less subject to genetic variability. There is a tendency to use lesser amounts of salts and hormones in media to reduce or prevent callus formation and to bring about a better quality of growth in many instances. An ever increasing number of plantlets are bypassing Stage III (rooting medium). Plantlets are removed from State II directly to soil, with or without auxin treatment, and held under high humidity until they have rooted. Only experience will indicate if this step is efficient for your particular cultivar and operation. When Stage III is not bypassed charcoal can usually be added to Stage III medium to help promote rooting.

Cleaning of the explants may be done in the media preparation room through final mixing in bleach solution. They should then be taken to the sterile hood, still in bleach, and the final rinses applied, using sterile technique.

The organization of plant families in this culture guide is according to taxanomic order; thus the families that are more closely related are placed nearer to each other. Within any one family there are often similar tissue culture requirements or problems; these factors may not extend to nearby families.

The Murashige and Skoog (MS) Formula

Because the formula by Murashige and Skoog (MS) is used so frequently, it will not be itemized in detail for each of the species to follow. When MS is required, it will be indicated as MS salts, which includes MS major salts, MS minor salts, and iron ($FeSO_4$ and Na_2EDTA). The MS formula appears below. The procedures for stock solutions are found on pages 00–00. If a formula calls for 4628 mg/L of MS salts, you will need 4530 mg major salts, 33 mg minor salts, and 65 mg iron.

MS SALTS		*mg/L*	
Major salts	NH_4NO_3	1650	
	KNO_3	1900	
	$CaC1_2 \cdot 2H_2O$	440	
	$MgSO_4 \cdot 7H_2O$	370	
	KH_2PO_4	170	4530
Minor Salts	H_3BO_3	6.2	
	$MnSO_4 \cdot H_2O$	16.8	
	$ZnSO_4 \cdot 7H_2O$	8.6	
	KI	0.83	
	$Na_2MoO_4 \cdot 2H_2O$	0.25	
	$CuSO_4 \cdot 5H_2O$	0.025	
	$CoC1_2 \cdot 6H_2O$	0.025	32.73
Iron	Na_2EDTA	37.3	
	$FeSO_4 \cdot 7H_2O$	27.8	
			65.1
		Total mg	4627.83

References: #111, #113.

Alphabetical List by Genus

BOSTON FERN, *Nephrolepsis exaltata* (L.) Schott var. *bostoniensis* Davenport. Polypodiaceae.

Explant: 2.5 cm segments of runner tips.

Treatment: Remove 10 cm of actively growing fern runner tips. Cut into 2.5 cm pieces. Stir in 1/10 bleach for 15 minutes. Dip in 78% alcohol. Rinse in 3 sterile distilled-water rinses. In the hood cut into 5–10 mm pieces. Place on agar or in liquid medium.

Media: ½ or ¾ MS with kinetin (1.0 mg/L), and NAA (0.1 mg/L). Kinetin is omitted for Stage III.

Light: 100–300 f.c. from fluorescent light with 16 hrs light/8 hrs dark.

Temperature: 25°C.

Discussion: Pieces that grow in Stage I are transferred to the same medium for Stage II or they can be literally chopped up (after removal of the larger fronds) and spread on agar medium of the same composition. The latter procedure is a time saving process for mass production. Probably more Boston ferns are propagated by tissue culture than any other ornamental crop. With a possible production period of less than four months, a million plants per year is within reach of a modest lab.

References: #32, #88, #113, #117, #122

BOSTON FERN MEDIA

	Stage I & II	III
Compound	*mg/liter*	
MS salts	3086 (= ¾ strength)	3086
NaH_2PO_4	125	12.5
Inositol	500	500
Thiamine	5.0	5.0
Kinetin	1.0	—
NAA	0.1	0.1
Sucrose	20000	20000
Agar	8000	8000
pH 5.7		

STAGHORN FERN, *Platycerium stemaria.* Polypodiaceae.

Explant: 3 mm shoot apex from offshoot.

Treatment: Obtain offshoots with fronds less than 5 cm wide. Wash in running water for 5 minutes. Remove larger fronds and root-mass and discard. Excise 1 cm shoot tips and mix in 1/10 bleach for 10 minutes. Rinse in 3 rinses of sterile distilled water. Remove hairs. Mix in 1/20 bleach for 5 minutes. Excise 3 mm dome. Rinse in sterile distilled water.

Media: Modified MS with 15 mg/L IAA for Stages I and II. Omit IAA in Stage III. Lily medium also gives excellent results.

Light: 100–300 f.c. from cool white fluorescent light with 14 hrs light/10 hrs dark.

Temperature: 25°C.

Discussion: The explants may turn black overnight but growth can be expected in 6 weeks. Subdivide and transfer in 2 months. Make further transfers at 3-week intervals. A time saving practice for commercial production involves homogenizing Stage III cultures. Approximately 40 Stage III plantlets are placed in a sterile blender with 50 ml of sterile distilled water. The blender is turned on at low speed for 5 seconds. 10 ml portions are aseptically pipeted from the blender onto sterile media in culture dishes, jars, or flasks. In 2 months the rooted plantlets (possible 200) are transferred to potting mix or tree fern/sphagnum and hardened off under mist or plastic cover. Because some of the more desirable species, such as *P. wandae,* produce very few offshoots for vegetative propagation, they are good candidates for tissue culture.

References: #33, #73, #113.

STAGHORN FERN MEDIA

	Stage I & II	III
Compound	*mg/liter*	
MS salts	4628	4628
Nicotinic acid	1.0	1.0
Pyridoxine HC1	1.0	1.0
Thiamine HC1	.4	.4
$NaH_2PO_4 \cdot H_2O$	170	170
Adenine SO_4	80	—
IAA	15	—
Sucrose	30000	30000
Agar	8000	8000
pH 5.7		

LONGLEAF PINE *Pinus palustris*. Pinaceae.

Explant: Embryos from viable seeds.

Treatment: Agitate seed in 50% bleach for 15 minutes. Rinse three times in sterile water. Soak for 40 hours in sterile, distilled water. Rinse in 33% bleach for 5 seconds. Rinse in sterile distilled water. Dissect out firm, white, embryo.

Media: Low salts and low sucrose prevail in these media. Some roots will form in stage II medium; if not, try Stage III medium or try 10 mg/L IBA added to Medium II for 4 weeks then back to II without the IBA.

Light: 800–1000 f.c. from cool white fluorescent light with 16 hrs light/8 hrs dark.

Temperature: 25°C.

Discussion: More than a dozen pine species have been reported as having been tissue cultured but there is no agreement on media formulas. Many use MS in dilute or other modified form. Rooting and hardening off are especially difficult. Requirements seem to vary not only with the species but within species, and from clone to clone as well. Because of these problems *Pinus* and other conifer genera remain one of the most challenging groups of plants to micropropagate. (See page 00.)

References: #1, #19, #138, #150, #159.

LONGLEAF PINE MEDIA

Compound	Stage I	II (mg/liter)	III
$(NH_4)_2SO_4$	200	—	—
$CaCl_2 \cdot 2H_2O$	150	—	—
$Ca(NO_3)_2 \cdot 4H_2O$	—	300	—
$MgSO_4 \cdot 7H_2O$	250	740	—
KNO_3	1000	80	—
KCl	300	65	—
$NaH_2PO_4 \cdot H_2O$	170	170	—
Na_2SO_4	—	200	—
$FeSO_4 \cdot 7H_2O$	27.8	27.8	13.9
Na_2EDTA	37.3	37.3	18.7
MS minor salts	32.7	32.7	16.4
MS major salts	—	—	2265
Inositol	10	10	—
Thiamine HCl	1.0	1.0	—
Nicotinic acid	0.1	0.1	—
Pyridoxine HCl	0.1	0.1	—
NAA	2.0	—	1.0
IBA	—	—	1.0
BA	5.0	—	—
Sucrose	20000	20000	10000
Agar	7000	7000	7000
pH 5.7			

SEQUOIA SEMPERVIRENS (Coastal Redwood). Pinaceae.

Explant: 1–3 cm shoot tips.

Treatment: Wash in water with liquid detergent. Rinse. Rinse in 78% alcohol for 1 minute. Mix in tap water with 0.1% Tween 20 and 0.1% Captan (fungicide) for 20 minutes. Rinse. Mix in 1/10 bleach with 0.1% Tween 20 for 20 minutes. Rinse in 1/100 bleach followed by 78% alcohol for 10 seconds. Rinse three times in sterile distilled water.

Media: MS salts supplemented with sodium phosphate, inositol, adenine sulfate, thiamine, kinetin, and a trace of IAA suffice for starting and multiplication. Rooting will occur when transferred to ½ MS, ½ vitamins, lower sugar, no kinetin, both IAA and IBA as auxins, and charcoal. Gelrite is satisfactory for both multiplication and rooting.

Light: 100–300 f.c. from cool white fluorescent lamps with 16 hrs light,/8 hrs dark. After 4 weeks in rooting, a 2-week period in darkness may help.

Temperature: 25°C.

Discussion: Sequoia is one of very few conifers that lend themselves to culture with comparatively few problems. Cleaning and growing-on remain the most critical steps. Boulay suggests removal of shoots (4 cm) from rooting media cutting them just above the agar, soaking in commercial rooting hormone with benlate (125 mg/L) for 24 hours. He then plants them in perlite/vermiculite 4/1.

References: #16, #24, #113.

SEQUOIA SEMPERVIRENS MEDIA

Compound	Stage I & II	Stage III
	mg/liter	
MS salts	4628	2314
NaH_2PO_4	160	—
Inositol	100	50
Thiamine HCl	0.4	0.2
Adenine sulfate	80	—
Kinetin	2.0	—
IAA	0.5	2.0
IBA	—	3.0
Sucrose	30000	20000
Charcoal	am	600
Agar (Gelrite)	2000	2000
pH 5.6		

RYEGRASS × TALL FESCUE, *Lolium multiflorum* Lam. × *Festuca arundinacea* Schreb. Gramineae.

Explant: Midveins of leaves and lower ends of internodes and peduncles (flower cluster stalks), from plants with emerging panicles (inflorescences).

Treatment: Dip stalks in 70% ethyl alcohol for 10 seconds. Mix in 1/10 bleach for 15 minutes. Rinse in 3 rinses of sterile, distilled water for 1 minute each. Remove leaf sheath. Cut peduncle and internode sections into 3 mm segments.

Media: First, produce callus on modified MS with 2 mg/L of 2,4-D and 0.1 mg/L of NAA. Subculture callus every 3–4 weeks. After three subcultures, wait for 8 weeks and transfer to differentiation medium: reduce sugar to 10 g/L, agar to 5 g/L, and 2,4-D to 0.25 mg/L.

Light: 100–300 f.c. from cool white fluorescent lamps with 16 hrs light/8 hrs dark.

Temperature: 25°C.

Discussion: This particular hybrid is infertile, therefore completely dependent upon vegetative propagation for reproduction. Grasses and legumes have not lent themselves readily to culture in vitro. Fortunately, they grow well from seed. However, there are special instances, as the example given here, where tissue culture is particularly useful for multiplication, not to mention other applications, such as haploid culture, disease resistance, mutation inducement, etc.

References: #37, #77, #113, #154.

RYE × FESCUE MEDIA

Compound	Stage I	II & III
	mg/liter	
MS salts	4628	4628
Nicotinic acid	0.5	—
Pyridoxine HCl	0.5	—
Thiamine HCl	0.1	—
Glycine	2.0	2.0
2,4-D	2.0	0.25
NAA	0.1	—
Sucrose	30000	10000
Agar	10000	5000
pH 5.7		

ANTHURIUM ANDREANUM Lind. Araceae.

Explant: Vegetative buds.

Treatment: Cut nodal sections of approximately 1 × 1 × 0.5 cm, each containing a bud. Soak for 20 minutes in 1/10 bleach with 0.1% Tween 20. Under a microscope remove leaf coverings and excise buds no larger than 2 mm at their base. Soak buds in 1/10 bleach for 30–45 minutes. Rinse in sterile distilled water for 5 minutes.

Media: For Stage I 15% coconut milk is added to MS salts plus thiamine, nicotinic acid, pyridoxine and 20% sucrose. Buds are rotated in this liquid medium for 6 weeks. For Stage II transfer 2-node shoot sections to the same medium but without coconut milk and with agar and 0.2 mg/L BA.

Light: Continuous illumination at 100 f.c. from fluorescent lights.

Temperature: 25°–28°C.

Discussion: The cultivars "Mauna Kea," "Calypso," and "Trinidad" are among the varieties successfully cultivated by this method. The disinfecting process described is the most successful in combatting a persistant contamination problem identified with this species.

References: #80, #113, #123.

ANTHURIUM MEDIA

	Stage I	II	III
Compound		*mg/liter*	
MS salts	4628	4628	4628
Inositol	100	100	100
Nicotinic acid	0.5	0.5	0.5
Pyridoxine HCl	0.5	0.5	0.5
Thiamine HCl	0.4	0.4	0.4
BA	—	0.2	—
Coconut milk	150 ml	—	—
Sucrose	20000	20000	20000
Agar		8000	8000
pH 5.5			

ARROWHEAD VINE, *Syngonium* spp. Araceae.

Explant: Shoot tips 1–5 mm.

Treatment: Obtain 1–2 cm shoot tips from young plants. Mix for 15 minutes in 1/10 bleach with 0.1% Tween 20. Rinse well in 3 rinses of sterile, distilled water. Excise 1–5 mm of shoot apex.

Media: Stage I and II are carried out in stationary or agitated liquid MS salts, sodium phosphate, vitamins, 2iP, and IAA. For Stage III use MS salts and agar.

Light: 300 f.c. from cool white fluorescent lights for Stage I and II. 800 f.c. for Stage III. Stages I–III 16 hrs light/8 hrs dark.

Temperature: 25°C.

Discussion: This desirable houseplant produces interesting variations which have the potential of economical multiplication through tissue culture.

References: #89, #113.

ARROWHEAD VINE MEDIA

	Stage I & II	III
Compound	*mg/liter*	
MS salts	4628	4628
$NaH_2PO_4 \cdot H_2O$	340	—
Inosital	100	—
Thiamine HCl	400	—
2iP	3.0	—
IAA	1.0	—
Sucrose	30000	—
Agar	—	8000
pH 5.5		

ASPARAGUS OFFICINALIS L. Liliaceae.

Explant: Lateral buds.

Treatment: Agitate spears, 15–20 cm long, in 1/10 bleach for 10 minutes. Wash in distilled water for 5 minutes. Excise lateral buds.

Media: Modified MS with NAA (1.0 mg/L), kinetin (1.0 mg/L), and vitamins.

Light: 100 f.c. from cool white fluorescent light for 16 hr light/8 hr dark.

Temperature: 29°C.

Discussion: Cut and transfer single bud sections from the spears which develop in 6 weeks.

Normally, asparagus from seed is too varied for optimum commercial production and multiplication by division is very labor intensive. Tissue culture of selected plants is a viable alternative. Micropropagation of parent lines for production of seeds of known background is a particularly valuable, practical application.

References: #29, #67, #110, #113, #126, #170.

ASPARAGUS MEDIA

Compound	Stage I & II	III
	mg/liter	
NH_4NO_3	1320	1320
KNO_3	1520	1520
$MgSO_4 \cdot 7H_2O$	370	370
$CaCl_2 \cdot 2H_2O$	440	440
KH_2PO_4	170	170
MS minor salts	32.7	32.7
$FeSO_4 \cdot 7H_2O$	27.3	27.3
Na_2EDTA	37.3	37.3
Pyradoxine HCl	0.5	0.5
Thiamine HCl	0.1	0.1
Kinetin	1.0	0.1
NAA	1.0	0.1
Sucrose	25000	25000
Agar	7000	6000

HEMEROCALLIS spp. (Daylily). Liliaceae.

Explant: 2 mm sections of young (10 cm) inflorescence scapes. Flower petals and sepals from 1.0 mm-long flower buds.

Treatment: Remove and discard bract tissue from flower buds. Wash inflorescence in water with 0.1% Tween 20. Rinse. Mix in 10% bleach with Tween 20 for 20 minutes. Rinse 3 times in sterile water. In the hood moisten sterile toweling with sterile antioxidant (p. 00). Slice scape into 2 mm sections. Place sections upside down on medium in test tube.

Media: Modified MS with KH_2PO_4, casein hydrolysate, malt extract, and adenine sulfate. Hormones are NAA at 10 mg/L and kinetin at 0.1 mg/L for callus formation with the NAA lowered to 0.5 for plantlet formation.

Light: Continuous dark for 4–8 weeks for callus followed by 300–1000 f.c. continuous light for plantlet development (2–6 months).

Temperature: 26°C.

Discussion: Daylily meristems are not used as explants because they are difficult to clean and the plant is destroyed if they are removed. Transplant plantlets to 2 sand:1 loam:1 peat:1 perlite. Water with Hoagland's mist for 10 days.

References: #42, #60, #61, #87.

DAY LILY MEDIA

Compound	Callus	Plantlets
NH_4NO_3	1650	1650
KNO_3	1900	1900
$CaCl_2 \cdot 2H_2O$	440	440
KH_2PO_4	300	300
$MgSO_4 \cdot 7H_2O$	370	370
MS minor salts	32.8	32.8
$FeSO_4 \cdot 7H_2O$	27.8	27.8
Na_2EDTA	37.3	37.3
Inositol	100	100
Thiamine HCl	0.4	0.4
Kinetin	0.1	0.1
$AdSO_4$	160	160
NAA	10	0.5
Malt extract	500	500
Casein hydrolysate	500	500
Sucrose	60000	30000
Agar	6000	6000

pH 5.5

HOSTA spp. Liliaceae.

Explant: Flower buds, .5–1 cm long; and scape (flower stem) sections, 3 mm long.

Treatment: Cut inflorescence, including about 6 cm of scape. Remove buds and cut scape into 3 pieces. Immerse buds and scape pieces into 70% ethyl alcohol for 30 sec. Rinse in sterile distilled water for 2 minutes. Mix in 1/10 bleach with 0.1% Tween 20 for 10 minutes. Rinse well in 3 rinses of sterile, distilled water. Cut scape pieces into 3 mm sections.

Media: Modified MS with vitamins, NAA (0.5 mg/L) and BA (2 mg/L). Meyer added adenine sulfate and casein hydrolystate for bud culture in standard light conditions. Reduce BA to 0.1 mg/L for Stage II and III.

Light: Some varieties initiate shoots after 4 weeks of darkness (*H. lancifolia, H. subcordata grandiflora* 'Royal Standard,' *H. 'Honeybells,' and H.* 'Aoki'), followed by 100–300 f.c. from fluorescent lights with 16 hrs light/8 hrs dark. Some species initiate shoots without a period of continuous darkness (*H. sieboldiana* flower buds). Still others produce callus in the dark followed by shoot growth in standard light conditions.

Temperature: 26°C.

Discussion: There is room for considerable experimentation with *Hosta* species and cultivars. The fact that some regenerate directly in light conditions, and that others require darkness for callus production and/or shoot initiation suggests that plant readiness versus hormone balance needs further study.

References: #99, #113, #119, #120, #157, #173, #174.

HOSTA MEDIA

Compound	Stage I	II	III
		mg/liter	
NH_4NO_3	1650	1650	1650
KNO_3	1900	1900	1900
$CaCl_2 \cdot 2H_2O$	440	440	440
KH_2PO_4	300	300	300
$MgSO_4 \cdot 7H_2O$	370	370	370
$NaH_2PO_4 \cdot H_2O$	170	170	170
Ms minor salts	32.7	32.7	32.7
$FeSO_4 \cdot 7H_2O$	27.8	27.8	27.8
Na_2EDTA	37.3	37.3	37.3
Inositol	100	100	100
Thiamine HCl	0.4	0.4	0.4
Glycine	2.0	2.0	—
Adenine sulfate	160	160	—
NAA	0.5	0.5	0.5
BA	2.0	0.1	0.1
Casein hydrolysate	500	500	500
Sucrose	30000	30000	30000
Agar	6000	6000	6000
pH 5.5			

LILY, *Lilium* spp. Liliaceae.

Explant: 4–5 mm cubes from bulb scale bases. Young axillary buds, young (¼–½") flower buds, young shoots, leaf bases, axillary bulbils, and ovary cross sections (1–3 mm) of mature flower buds (unpollinated) have also been successful.

Treatment: Separate inner bulb scales such that they retain a portion of the baseplate. Wash in tap water with few drops of detergent. Rinse in running tap water for 15 minutes. Rinse briefly in sterile water. Mix in 1/10 bleach for 10–20 minutes. Rinse in 3 rinses of sterile distilled water for one minute each. Cut 4–5 mm cubes from base of scales including some basal plate tissue.

Media: Asiatic hybrids and orientals usually respond well to MS with ½ strength ammonium nitrate and a minimum of amendments with 0.03 mg/L NAA. 2iP-IAA formulations are strong and more designed for Easter lilies. For bud culture use MS plus organics with 0.1 mg/L BA and 0.1 mg/L NAA or 1.0 mg/L BA and 1.0 mg/L NAA.

Light: Both intermittent light and continuous darkness are used depending on the operator and the cultivar. Continuous darkness is preferred as standard as it leads toward more bulblets and fewer leaves (which is better for transplanting). Bud cultures are better under 100–300 f.c. from fluorescent lights with 16 hr light/8 hr dark.

Temperature: 21°C–30°C with warmer better than cooler if no more than 30°C. Chilling temperatures are 1.1–8.9°C; 4–6 weeks for asiatics and 6–10 weeks for orientals.

Discussion: Most commercially micropropagated lilies are the Asiatic hybrids and orientals. Cultures can be chilled in vitro before planting but most growers find it useful to rinse, put into loose, damp medium (vermiculite), then chill. Anderson cultured *Lilium,* cv. 'Red Carpet' from scale base sections. Cultures were grown at 100 f.c. with 16 hrs light/8 hrs dark at 21°C, followed by planting into Jiffy 7's. These were covered with glass under 300–600 f.c. of light and grown to 1-gram size before chilling. Stimart, et al, cultured Easter lilies, *Lilium longivlorum* 'Ace' and 'Nellie White', from scale base sections on Linsmaier-Skoog medium supplemented with 0.03 mg/L NAA. Cultures were grown in continuous darkness at 25°C followed by 3 weeks at 4°C before planting into vermiculite.

References: #8, #12, #74, #94, #113, #142, #143, #147, #148, #157.

LILY MEDIA

Compound	Stage I & II	III (Optional)
	mg/liter	
NH_4NO_3	850	850
KNO_3	1900	1900
$CaCl_2 \cdot 2H_2O$	440	440
$MgSO_4 \cdot 7H_2O$	370	370
KH_2PO_4	170	170
MS Minor Salts	32.7	32.7
Na_2EDTA	37.3	37.3
$FeSO_4 \cdot 7H_2O$	27.8	27.8
Inositol	100	100
Thiamine	0.4	0.4
Adenine sulfate	80	—
NAA	0.03	0.03
Sucrose	30000	30000
Agar	8000	8000
pH 5.7		

AMARYLLIS, *Hippeastrum hybridum.* Amaryllidaceae.

Explant: Sections of bulb scales, ovary tissue, pedicels, young flower buds (grown 10 cm above bulb neck).

Treatment: Wash bulbs in water with 0.1% Tween 20 followed by 8 minutes in 20% bleach. Rinse 3 times in sterile water. Cut 1 cm squares from inner scales. Wash in 10% bleach for 2 minutes. Rinse well. Wash stems and tight flower buds in water with 0.1% Tween 20. Rinse for 20 minutes in sterile water. Dip in 90% alcohol for 10 seconds. Rinse for 1 minute in sterile water. Mix for 15 minutes in 10% bleach or 8% calcium hypochlorite. Remove spathe valves from buds. Slice pedicels 3 mm thick and invert on medium.

Media: Modified MS with additives depending upon the reference.

Light: Six weeks in dark followed by 12 hrs light/12 hrs dark.

Temperature: 25°C.

References: #74, #102, #113, #134, #157, #166.

AMARYLLIS MEDIA

	Stage I & II		III
	Scales	Other Explants	
Compound		*mg/liter*	
MS salts	4628	—	4628
NH_4NO_3	—	1630	—
KNO_3	—	1900	—
$CaCl_2 \cdot 2H_2O$	—	440	—
$MgSO_4 \cdot 7H_2O$	—	370	—
$KH_2PO_4 \cdot H_2O$	—	300	—
$FeSO_4 \cdot 7H_2O$	—	27.8	—
Na_2EDTA	—	37.3	—
MS minor salts	—	32.7	—
Inositol	100	100	—
Nicotinic acid	0.5	1.0	—
Pyridoxine HCl	0.5	1.0	—
Thiamine HCl	0.1	1.0	—
Glycine	2.0	—	—
Kinetin	0.5	0.5	—
$AdSO_4$	—	2.5	—
NAA	5.0	2.5	—
Casein hydrolysate	1000	500	500
Malt extract	—	500	500
Sucrose	30000	30000	30000
Agar	8000	6000	6000
pH 5.5			

STAR GRASS *Hypoxis rooperi* S. Moore, Amaryllidaceae

Explant: Upper half of corm.

Treatment: Remove and discard leaves and roots from corm. Cut upper half of corm horizontally into three slices. Remove outer layer (epidermis) of tissue. Wash in running water for 20 min. Stir in 100% ethyl alcohol for 5 min. Soak in 15 benomyl (Benlate) for 15 min. Stir in 0.1% mercuric chloride or 1/10 bleach for 15 min. Rinse well in sterile water.

Media: MS with 1 mg/L BA. Sucrose, inositol, thiamine and 1000 mg/L casein hydrolysate are other additives. The same medium without hormones is used for rooting.

Light: 100–300 f.c. cool white fluorescent light with 16 hrs light/8 hrs dark.

Temperature: 25°C.

Discussion: This is a species from Africa

related to our delightful rockery species, H. hirsuta. H. rooperi has been used for centuries in Africa as food and as a medicinal. The medicinal qualities of Hypoxis spp. have recently been recognized in the U.S. (See K. H. Pegel, 1980, "Active plant extracts of Hypoxidaceae and their use. U.S. Patent No. 4198401.). The techniques well might apply to other corms.

References: #113, #118, #157.

STAR GRASS MEDIA

Compound	Stage I & II	Stage III
	mg/liter	
MS salts	4628	4628
Thiamine HCl	1.0	1.0
Inositol	100	100
Casein hydrolysate	1000	1000
BA	1.0	—
Sucrose	30000	30000
Agar	8000	8000
pH 5.8		

FREESIA spp. Iridaceae.

Explant: 2 mm transverse slices of aerial corms.

Treatment: Agitate aerial corms in 95% ethyl alcohol for 2 minutes. Mix in 1/10 bleach with 0.1% Tween 20 for 10 minutes. Rinse in 3 changes of sterile, distilled water for 2 minutes each. Slice 2 mm cross sections.

Medium: Modified MS with NAA at 0.1 mg/L and kinetin at 0.05 mg/L promotes both shoots and roots.

Light: Continuous dark until formation of shoots and roots, then place in 100 f.c. from fluorescent lights with 16 hrs light/8 hrs dark.

Temperature: 25°C.

Discussion: Growth in the dark occurs in about 4 weeks. The light period need only be about 2 weeks after which the plantlets can be grown-on in moist peat under fluorescent lights at 20°C. Water with MS nutrient solution for a week, then with tap water. Plant in soil in greenhouse when 8 weeks out of culture.

References: #98, #113, #126.

FREESIA MEDIUM

Compound	Stage I, II, & III
	mg/liter
MS salts	4628
Inositol	100
NAA	0.1
Kinetin	.05
Sucrose	30000
Agar	8000
pH 5.8	

IRIS spp. (Bearded) Iridaceae

Explant: 2 mm sections of peduncle (flower stem) of young inflorescences.

Treatment: With a sterilized knife cut young, 15 cm (6"), inflorescence. Remove and discard bract tissue. Stir inflorescence in 1/10 bleach with 0.1% Tween 20 for 20 minutes. Rinse in three changes of sterile water for 2 minutes each. Have ready antioxidant solution (0.1% citric acid/0.1% ascorbic acid solution). Remove florets and place on sterile paper toweling moistened with antioxidant solution. Slice 2 mm sections of peduncles. Place upside down on agar medium.

Medium: Modified MS with KH_2PO_4 (300 mg/L), casein hydrolysate (500 mg/L), malt extract (500 mg/L), and adenine sulfate (160 mg/L). Use 2.5 mg/L NAA and 0.5 mg/L kinetin for callus formation.

Light: Continuous darkness for 6–12 weeks followed by 100–300 f.c. fluorescent light with 16 hrs light/8 hrs dark.

Temperature: 26°C.

Discussion: The callus that develops from the explant in 6–12 weeks is cut into several pieces and placed in the light. Shoots and roots develop from the edges of the callus pieces in

2–5 months. Transplant to 1 peat/1 perlite for 2 weeks under mist.

References: #73, #74, #98, #113.

BEARDED IRIS MEDIUM

Compound	Stage I & II mg/liter
NH_4NO_3	1650
KNO_3	1900
$CaCl_2 \cdot 2H_2O$	440
KH_2PO_4	300
$MgSO_4 \cdot 7H_2O$	370
MS minor salts	32.8
$FeSO_4 \cdot 7H_2O$	27.8
Na_2EDTA	37.3
Inositol	100
Thiamine HCl	0.4
Kinetin	0.5
Adenine sulfate	160
NAA	2.5
Malt extract	500
Casein hydrolysate	500
Sucrose	30000
Agar	6000
pH 5.5	

IRIS XIPHIUM (Bulbous). Iridaceae.

Explant: In September trim away all but the basal plate with the tiny sheath leaves covering the shoot tip which has started to grow (.5–1.5 mm). Cut away and discard the basal plate saving only the shoot tip.

Treatment: Mix in sterile water with 0.1% Tween 20 for 10 minutes. Rinse. Mix in 10% bleach for 15 minutes. Rinse thoroughly in sterile water.

Medium: 3 stages of medium are used: multiplication, intermediate, and bulbing. The first is modified MS with BA and IAA; the second omits $AdSO_4$, BA, and IAA, but adds kinetin and NAA, and increases the sucrose. The bulbing stage medium omits the hormones.

Light: Multiplication and intermediate stage: 16 hrs light/8 hrs dark. Bulbing stage: Continuous dark for 6 months.

Temperature: Multiplication and intermediate stage: 20°C. Bulbing stage: 25°–28°C.

Discussion: The presence of virus significantly lowers the quality of bulbs. Because virus infection plague the iris bulb industry, "disease-free" bulbs bring premium prices. The first transfer into multiplication is at about 6 weeks. Two shorter transfer periods should see 1–2 mm buds form. At this stage it is desirable to have a commercial plant pathology laboratory determine by electron microscopy if virus are present. Transfer the individual 1–2 mm buds to multiplication medium. A single cycle of 3–5 weeks in the intermediate medium is all that is required for conditioning for bulbing. Following 6 months in bulbing, plant in a cool (10°C) greenhouse for rooting.

References: #5, #101, #113.

BULBOUS IRIS MEDIUM

Compound	Multi.	Inter. mg/liter	Bulbing
Ms salts	4628	4628	4628
Inositol	100	100	100
Thiamine HCl	0.4	0.4	0.4
Adenine sulfate	80	—	—
IAA	1.0	—	—
NAA	—	0.1	—
BA	1.5	—	—
Kinetin	—	0.01	—
Sucrose	30000	60000	60000
Agar	6000	6000	6000
pH 5.7.			

CATTLEYA spp. Orchidaceae.

Explant: Axillary buds or meristems from 2 to 5 cm shoots.

Treatment: Wash the shoot in water with detergent. Dip the shoot in alcohol for 10 seconds. Mix in 1/10 bleach for 15 minutes. Rinse briefly in sterile water and dry on sterile paper towel. Under a dissecting microscope, carefully remove the overlapping leaves. To help prevent browning dissect under fresh sterile antioxidant solution. Watch for buds about 2 mm in size at the base of the leaves. Excise the whole buds, severing just below the point of attachment. These may be cultured, or dissect further to obtain meristem with one or two pairs of leaf primordia. See details on strawberry meristemming, page 120.

Media: Grow buds or meristems in medium with 100 ml/L coconut milk. They are best started in agitated liquid but agar solidified medium is often satisfactory, and required for Stage III.

Light: Continuous light at 100–300 f.c. from cool white fluorescent lamps.

Temperature: 26°C.

Discussion: Cattleyas multiply better in liquid than on agar, at least initially. The theoretical reason is that the agitation inhibits polarity (orientation). Once polarity is established, the cultures put out shoots and roots and mature. The initial growth that is desired for multiplication is a mass of protocorms. As soon as this mass grows to 1 cm it should be divided and put back into liquid or agar medium. When the culture is on agar frequent division will help upset orientation and delay plantlet formation.

References: #15, #64, #108, #109, #113, #129, #167.

CATTLEYA ORCHID MEDIUM

See *Cymbidium* orchid medium, modified Knudson's "C" formula.

CYMBIDIUM. spp. Orchidaceae.

Explant: Apical meristem.

Treatment: Remove outer leaves from 3 cm shoots. Dip in ethyl alcohol 2 seconds then mix in 1/10 bleach for 15 minutes. Rinse in sterile water. Remove remaining leaves. Excise meristem consisting of apical dome, two leaf primordia, and a cube of tissue, all less than .5 mm.

Media: Cymbidiums will grow well in either liquid or agar medium but liquid treatment may produce faster and more abundant growth. Start in a modified Knudson "C" medium (or Vacin and Went, see pages 138 and 139). After three or four transfers, place on agar medium (Knudson "C" or Morel and Muller, pages 138, 139) for shoot and root development.

Light: Continuous light of 100 f.c. from fluorescent lights.

Temperature: 22°C.

Discussion: When orchid seeds germinate they first produce protocorms, bulbous tissue with root hairs. Apical meristems in culture follow a similar development. They should show greening within a week and protocorm masses should be ready for division and transfer in two months, or when they have grown to half a centimeter. To obtain coconut milk, purchase a coconut from the store. Shake it before buying to be sure there is liquid in it—you can hear it sloshing. Pound two holes in the eyes with a nail. Pour out the liquid. Coconut "water" is available from some suppliers (GIBCO).

Vacin and Went formula plus coconut milk (coconut water) has also been used for cymbidiums as well as dendrobiums and phalaenopsis.

References: #15, #108, #109, #129, #167.

CYMBIDIUM ORCHID MEDIUM

Modified Knudson's "C"

Compound	Stage I, II, & III *mg/liter*
$Ca(NO_3)_2 \cdot 4H_2O$	1000
$(NH_4)_2SO_4$	500
$MgSO_4 \cdot 7H_2O$	250
KH_2PO_4	250
$FeSO_4 \cdot 7H_2O$	27.8
Na_2EDTA	37.3
MS minor salts	32.8
Coconut milk	100 ml
Sucrose	20000
Agar (when used)	6000
pH 5.5	

CARNATION, *Dianthus caryophyllus* L. Carophyllaceae.

Explant: 1 mm shoot tip.

Treatment: Remove vegetative stems with approximately 12 nodes from stock plants. Remove all leaves over 3 mm long. Wash in water with few drops of detergent. Mix in 1/10 bleach with 0.1% Tween 20 for 5 minutes. Rinse twice in sterile distilled water for 3 minutes. Aseptically meristem 1 mm stem tips with one or two pairs of leaf primordia and place on agar medium.

Media: Stage I: Modified MS salts with 2 mg/L kinetin, 0.2 mg/L NAA, inositol (50 mg/L), glycine (2 mg/L), and casein hydrolysate (3 g/L). Stage II: Omit casein hydrolysate and lower the kinetin to 0.5 mg/L, and the NAA to 0.1 mg/L. Stage III: Omit kinetin and NAA from medium, or root 2 cm shoots directly in Jiffy 7's.

Light: 200 f.c. from cool white fluorescent lamps with continuous light.

Temperature: 22°C.

Discussion: Carnations are commercially meristemmed to insure virus-free stock.

References: #35, #43, #67, #85, #113.

CARNATION MEDIA

Compound	Stage I	II *mg/liter*	III
MS salts	—	4628	4628
NH_4NO_3	400	—	—
KNO_3	80	—	—
KC1	65	—	—
$MgSO_4 \cdot 7H_2O$	72	—	—
KH_2PO_4	12.5	—	—
$Ca(NO_3)_2 \cdot 4H_2O$	144	—	—
$FeSO_4 \cdot 7H_2O$	27.8	—	—
Na_2EDTA	32.8	—	—
$MnSO_4 \cdot H_2O$	6.5	—	—
$ZnSO_4 \cdot 7H_2O$	2.7	—	—
H_3BO_3	1.6	—	—
KI	.75	—	—
Cinnamic acid	1.5	1.5	1.5

Compound	Stage I	II	III
		mg/liter	
D-Ca-Pantothenate	5.0	5.0	5.0
Inositol	50	100	100
Nicotinic acid	0.5	0.5	0.5
Pyridoxine HCl	0.5	0.5	0.5
Thiamine HCl	0.1	0.1	0.1
Glycine	2.0	2.0	2.0
Kinetin	2.0	0.5	—
NAA	0.2	0.1	—
Casein hydrolysate	3000	—	—
Sucrose	30000	30000	30000
Agar	6000	6000	6000
pH 5.0			

NANDINA DOMESTICA. Berberidaceae.

Explant: Lateral and terminal buds.

Treatment: Remove outer leaves from 5 cm shoot tips. Wash shoots in warm water with liquid detergent for 10 minutes. Rinse briefly with water. Agitate in 1/10 bleach with 0.1% Tween 20 for 20 minutes. On sterile paper toweling pieces or filter paper excise buds under a dissecting microscope.

Media: A modified Gamborg's B5 medium is used with MS iron for Stages I and II. One third MS salts is used for rooting. The major differences between B5 and MS are that B5 uses sodium phosphate instead of potassium phosphate and ammonium sulfate instead of ammonium nitrate. MS may be used instead of B5 but B5 is preferred. Uniquely, charcoal is used to help initiation.

Light: 600 f.c. from Sylvania Gro- Lux lamps with 16 hrs light/8 hrs dark. Gro-Lux lamps are not usually specified because cool white fluorescent lamps are commonly satisfactory. However, because two different studies of *Nandina* used Gro-Lux, they are suggested here.

Temperature: 26°C.

Discussion: J. Matsuyama's report on 'Royal Princess' and Roberta Smith's article on (dwarf) 'Purpurea' are two of few articles on this likely subject for tissue culture. Jean Gould and T. Murashige determined that the yellow exudate into culture media is due to the alkaloid berberine. As an alkaloid it is different from other cultures where the exudates are presumed to be phenolics.

References: #53, #90, #113.

NANDINA MEDIA

Compound	Stage I	Stage II	Stage III
		mg/liter	
$NaH_2PO_4 \cdot H_2O$	150	150	—
KNO_3	2500	2500	—
$(NH_4)_2SO_4$	134	134	—
$MgSO_4 \cdot 7H_2O$	250	250	—
$CaCl_2 \cdot 2H_2O$	150	150	—
$MnSO_4 \cdot H_2O$	10	10	—
H_3BO_3	3.0	3.0	—
$ZnSO_4 \cdot 7H_2O$	2.0	2.0	—
KI	.75	.75	—
$Na_2MoO_4 \cdot 2H_2O$	0.25	0.25	—
$CuSO_4 \cdot 5H_2O$	.025	.025	—
$CoCl_2 \cdot 6H_2O$	.025	.025	—
Na_2EDTA	37.3	37.3	—
$FeSO_4 \cdot 7H_2O$	27.8	27.8	—
MS salts			1543 (= ⅓ strength)
Thiamine HCl	0.4	0.4	0.4
Inositol	100	100	100
NAA	0.1	0.1	3.0
BA	1.0	1.0	—
Sucrose	30000	30000	30000
Charcoal	1500	—	1500
pH 5.7			

BROCCOLI, *Brassica oleracea* (Italica group). Cruciferae.

Explant: Flower buds (2 mm) and lateral shoot tips (1 cm).

Treatment: Precondition stock plants by holding in a greenhouse at 16°C for three weeks. Obtain young flower heads bearing 2 mm buds. Stir heads and lateral shoots in 1/10 bleach for 15 minutes. Excise 2 mm buds. Rinse buds and shoots in 1/100 bleach for 15 minutes.

Media: MS salts with $NaH_2PO_4 \cdot H_2O$, inositol, thiamine, and adenine sulfate. For Stage I and II add 1 mg/L of IAA and 4 mg/L of 2iP. For rooting, omit 2iP.

Light: For Stage I and II use 100 f.c. from cool white fluorescent lights with 16 hrs light/8 hrs dark. For Stage III increase intensity to 600 f.c.

Temperature: 19°C–23°C.

Discussion: Transplant rooted plantlets to 1 peat/1 perlite under intermittent mist for 3 days at which time humidity can be reduced. Plants are ready for transplant in 7 weeks. A major application of broccoli tissue culture is the multiplication of parents of a self-incompatible hybrid that is particularly desirable. The parents, multiplied by tissue culture, are planted in the field for open pollinated seed production.

References: #9, #10, #113.

BROCCOLI MEDIA

Compound	Stage I & II	III
	mg/liter	
MS salts	4628	4628
$NaH_2PO_4 \cdot H_2O$	170	170
Inositol	100	100
Thiamine HCl	0.4	0.4
Adenine sulfate	80	80
IAA	1.0	1.0
2iP	4.0	—
Sucrose	30000	30000
Agar	8000	8000
pH 5.7		

KALANCHOË, BOSSFELDIANA Poelin. Crassulaceae.

Explant: Leaf blade sections, 1½ cm × 1½ cm, stem sections, 2 cm, and shoot tips, 2 mm.

Treatment: Wash cuttings (3–5 cm) in warm water with detergent. Rinse briefly in distilled water. Mix in 1/10 bleach with 0.1% Tween 20 for 15 minutes. Rinse in 3 changes of sterile, distilled water for 1 minute each. Trim to size.

Media: Modified MS salts with 1 mg/L each of kinetin and IAA. For rooting, omit kinetin and IAA, and add 1 mg/L IBA.

Light: 100–300 f.c. from cool white fluorescent lights with 16 hrs light/8 hrs dark.

Discussion: With only 4–6 weeks from culture initiation to potted plantlets, this species is ideal for classroom tissue culture demonstration.

References: #113, #136.

KALANCHOË MEDIA

Compound	Stage I & II	III
	mg/liter	
MS salts	4628	4628
$NaH_2PO_4 \cdot H_2O$	120	120
Inositol	100	100
Thiamine HCl	0.4	0.4
Kinetin	1.0	—
Adenine sulfate	80	—
IAA	1.0	—
IBA	—	1.0
Sucrose	30000	30000
Agar	6000	6000
pH 5.7		

KIWIFRUIT, *Actinidia chinensis.* Rosaceae.

Explant: Meristems 0.2 to 0.5 mm long, or apical shoots 3 cm long.

Treatment: Dormant shoots taken in winter may be stored at 1°C (spray with fungicide) for later explanting. Wash shoots with water and Tween 20 for 20 minutes. Rinse, dip in 78% alcohol for 10 seconds, rinse. Mix in 1/10 bleach with 0.1% Tween 20 for 15 minutes. Rinse 3 times in sterile distilled water. Again, dip in alcohol for 10 seconds. Rinse twice in sterile distilled water. Dissect out the meristems, cleaning instruments frequently. Shoots may also be taken in late spring: Cut 3 cm shoot tips and remove explanded leaves. Stir in 6% bleach with 0.1% Tween 20 for 20 minutes. Rinse 3 times with sterile distilled water. Remove 1 cm from shoot base and place tip on agar medium.

Media: For initiation use ¾ strength MS with given additives. The sugar and inositol are slightly lower than normally prescribed. Agitated liquid medium for Stage II has provided excellent results, therefore is recommended here. Dispense into 125 ml flasks and cap with cotton and aluminum foil. Stage III may well be omitted by rooting Stage II shoots directly in soil mix; use of a preliminary dip into 2 mg/L IBA may be advantageous. Employing a Stage III auxin has caused excessive callus.

Light: 200–300 f.c. from fluorescent lamps with 16 hrs light/8 hrs dark.

Temperature: 23°C.

Discussion: Monette found that 2 bacterial contaminants that survived explant sterilization treatment did not pose a problem to kiwi proliferation or rooting.

The popularity of kiwifruit is increasing dramatically. Kiwi acreage has more than doubled in both Italy and California in the past 5 years. Because it is monoecious, only vegetative propagation is practical. Cuttings take 2 to 3 years to provide field plants while field plants are produced in one year through tissue culture. With these facts in mind it is probably a good time to start micropropagating choice field selections of kiwi as some labs are already doing.

References: #106, #113.

KIWIFRUIT MEDIA

	Stage I	Stage II	Stage III (Optional)
Compounds		*mg/liter*	
MS salts	3471 (= ¾ strength)	4628	2314 (= ½ strength)
$NaH_2PO_4 \cdot H_2O$	128	170	—
$AdSO_4$	60	80	—
Inositol	75	100	100
Thiamine HCl	0.3	0.4	0.4
BA	2.0	2.0	—
IBA	.023	0.03	—
Sucrose	22500	30000	20000
Agar	7000	(liquid)	—
pH	5.7	5.0	5.7

SASKATOON *Amelanchier alnifolia* 'Smoky.' Rosaceae.

Explant: Apical 2 cm of young, actively growing shoots.

Treatment: Remove all outer leaves. Dip in 78% alcohol. Stir in 7% bleach for 10 minutes. Dip in 78% alcohol. Rinse 3 times in sterile, distilled water. Trim to 5 mm of tip.

Media: Several simple modifications of MS have been used successfully. The one presented here has low BA content to help avoid vitrification.

Light: 100–300 f.c. from cool white fluorescent lights with 16 hrs/light, 8 hrs/dark for multiplication. Continuous of the same light for rooting and early acclimation.

Temperature: 22°C.

Discussion: This hardy ornamental shrub is growing in popularity as new cultivars appear in the trade. While not among the more difficult of woody plants to tissue culture it tends to suffer from vitrification and acclimation problems.

The astute work of R. E. Harris differs from the protocol outlined here. He used higher BA initially (3.0 mg/L) followed by an agitated liquid phase, without hormones, to condition shoots for rooting. Shoots were planted in 3/1 perlite/vermiculite, dampened with Rapid Grow and IBA (1 mg/L). They were covered with plastic and placed under cool white fluorescent lights with continuous lighting.

References: #59, #60, #113.

SASKATOON MEDIA

	Stage		
	I	II	III
Compound		*mg/liter*	
MS salts	3471 (= ¾ strength)	4628	4628
$NaH_2PO_4 \cdot H_2O$	—	100	—
Thiamine	0.4	0.4	0.4
Inositol	100	100	100
BA	1.0	1.0	—
IBA	—	—	1.0
GA_3	0.05	0.05	—
Sucrose	30000	30000	30000
Gelrite	3000	3000	3000
Charcoal	—	—	0.6
pH 5.7			

STRAWBERRY, *Fragaria* spp. Rosaceae.

Explant: Meristem (0.5 mm) from apical and lateral buds from young runner tips.

Treatment: Remove 2" runner tips from runners that have not yet leafed out. Wash in distilled water with 0.1% Tween 20 for 5 minutes. Stir in 1/10 bleach for 10 minutes. Stir in two rinses of sterile, distilled water for 10 seconds each. Place on sterile toweling in petri dish under a dissecting microscope. Excise meristems as described on page 00.

Media: Use MS salts with standard vitamins. Use 1 mg/L of BA for Stage I and II. Rooting can be directly into potting mix, or in Stage III with 1 mg IBA and charcoal.

Light: 100–300 f.c. from cool white fluorescent lights with 16 hrs light/8 hrs dark.

Temperature: 25°C.

Discussion: More strawberries are propagated by micropropagation than any other food crop. In Europe, two leaders, P. Boxus in Belgium and C. Damiano in Italy have demonstrated the feasibility of mass producing virus-free strawberry plants by tissue culture, especially for field motherplants used by strawberry plant growers. Several million plants per year are being produced by a dozen or so labs, mainly in Europe.

References: #11, #25, #81, #113.

STRAWBERRY MEDIA

	Stage I & II	III
Compound	*mg/liter*	
MS salts	4628	4628
Inositol	100	100
Nicotinic acid	0.5	0.5
Pyridoxine HCl	0.5	0.5
Thiamine HCl	0.4	0.4
BA	1.0	—
IBA	—	1.0
Sucrose	30000	30000
Agar	6000	6000
Charcoal	—	800
pH 5.7		

APPLE, *Malus* spp. Rosaceae.

Explant: One or two node stem pieces from actively growing shoots with lateral buds showing. Also semi-dormant buds.

Treatment: Prepare nodal sections as for cherry (*Prunus spp.*), or use semi-dormant buds: Collect twigs with the swollen, greener, axillary buds. Remove bud similar to a chip bud, i.e., with a sliver of stem on top and a basal section of tissue. Wash vigorously in tap water with 1% Tween 20 for 10 minutes. Stir in 1/10–1/5 bleach with 0.1% Tween 20 for 10–20 minutes, depending on hardness of the shoot. In the hood, rinse briefly in 3 rinses of sterile, distilled water. With frequent sterilization of instruments, remove outermost budscales and any burned tissue. Place upright in agar medium.

Media: For Stage I use half-strength MS salts, full-strength iron, inositol, and thiamine HCl, with no hormones. Stage II usually requires full-strength MS but BA/IBA content is tricky. Try BA at 1.0 mg/L and IBA at 0.1 and change to 0.25/0.01 if material becomes too succulent or remains short. In Stage III, return to half-strength MS but retain full-strength iron (Fe and EDTA), thiamine HCl and inositol, or root directly from Stage II into soil.

Light: 100–300 f.c. from cool white fluorescent lamps with 16 hrs light/8 hrs dark.

Temperature: 24°C–26°C.

Discussion: Apples are quite particular in their specific requirements which can vary in small detail from species to species, cultivar to cultivar, and even clone to clone. Some experimentation is usually necessary to determine the precise factors to promote optimum response. Important variables include:

Concentration of salts—1, ½, or ¼.

Type and concentration of auxin—IBA, IAA, NAA.

Etiolation in Stage II.

In preparing for rooting, etiolate the cultures by placing them in continuous dark (in cardboard boxes) in the growing room. After 2–3 weeks place them in the light for 1–2 weeks to green up before rooting.

Chilling requirement: (3–6 weeks at 3°C before hardening off).

Brand and concentration of agar: Start with Difco Bacto agar.

Addition of charcoal.

Length of time in Stage III.

Interval between transfers: Transfer at least every 3 weeks, more often if "bleeding," or if negative or no response. Transfer to Stage II medium as soon as explant starts growing.

References: #42, #83, #113, #146, #178.

APPLE MEDIA

Compound	Stage I	II	III
		mg/liter	
MS salts	2281 (= ½-strength MS less $FeSO_4$ and Na_2EDTA)	4628	2281 (= ½-strength MS less $FeSO_4$ and Na_2EDTA)
$FeSO_4$	27.8		27.8
Na_2EDTA	37.3		37.3
Inositol	208	208	208
Thiamine HCl	2.5	2.5	2.5
BA		1.0	
IBA		0.1	0.2
Sucrose	30000	30000	30000
Agar (Difco Bacto)	6000	7000	6000
pH 5.7			

PLUM, CHERRY, *Prunus* spp. Rosaceae.

Explant: One or 2-node stem pieces from actively growing shoots with lateral buds showing.

Treatment: Collect 7–10 cm (4") shoots from the first growth in spring. Remove and discard leaves, leaving one cm of petiole still attached to the stem. Wash with vigorous agitation in tap water with 1% Tween 20 for 10 minutes. Rinse in running tap water for 2 minutes. Stir in 1/10–1/5 bleach with 0.1% Tween 20 for 10–20 minutes—the harder the shoot, the stronger the bleach and/or the longer the time in bleach solution. In the hood, rinse 3 times for a few seconds in sterile distilled water. Cut one or two node pieces with a short piece of stem above and below the node. Be sure to sterilize your instruments frequently between cuts. Place lower stem stub into agar so that explant is upright and the bud is above the agar level.

Media: Half-strength MS salts is usually used throughout with no hormones in Stage I, however, BA at 0.5 to 1.1 mg/L will help lateral bud break on slow tips. For Stage II, add BA at 0.25–0.5 mg/L and IBA at 0.01 mg/L. Rooting can be directly into soil from Stage II or use Stage III medium, omitting BA and increasing IBA to 0.2 mg/L. Yeast extract at 1 g/L is sometimes helpful for cherries in Stage II.

Light: 100–300 f.c. from fluorescent lights with 16 hrs light/8 hrs dark.

Temperature: 25°C–27°C.

Discussion: Explant buds should break and grow in 3 weeks; promptly transfer to Stage II medium. Cultures should be transferred at least every three weeks.

References: #42, #83, #113, #146.

PLUM, CHERRY MEDIA

	Stage		
	I	II	III
Compound		*mg/liter*	
MS salts	2314 (= ½ strength)	2314	2314
Inositol	208	208	208
Thiamine HCl	2.5	2.5	2.5
BA		0.5–0.25	
IBA		0.01	0.2
Yeast extract		1.0	
Sucrose	30000	30000	30000
Agar (Bacto Difco)	6000	6000	6000

PEACH, *Prunus persica* L. Rosaceae.

Explant: Actively growing shoot tips, 1.5 cm long.

Treatment: Immerse explants in 15% bleach for 30 minutes. Rinse 4 times in sterile distilled water. Trim to 0.5 cm before placing on Gelrite agar.

Media ("AP" medium): Based on Gamborg's B5 this medium has less $MgSO_4$, double the $(NH_4)_2SO_4$, double the $MnSO_4 \cdot H_2O$, and lower vitamin content. BA is added at 6 mg/L and IBA at 0.1 mg/L. Prerooting treatment consists of 2 weeks in liquid medium containing 2× the multiplication formula. For rooting use ½ strength KNO_3 and ½ strength NH_4SO_4 in multiplication medium plus 0–1.0 mg/L IBA.

Light: 200 f.c. from cool white fluorescent and Gro-Lux tubes (1:1) for 15 hrs light/9 hrs dark for Stage I and II. Reduce light intensity to 140 f.c. for rooting.

Temperature: 26°C.

Discussion: Considerable work has been done on peach with minor deviations from conventional media and with only modest success. Almedhi and Parfitt appear to have digressed from the norm with significant Stage I and II success using both juvenile and mature explant material. They supported shoot growth and multiplication of 56 varieties using AP medium. Establishment in soil continues to be a problem. However, it may well be discovered that AP medium lends itself successfully to other genera.

References: #2, #44, #57, #104, #121, #145, #175.

PEACH MEDIA (AP medium)

Compound	Stage I & II	III
	mg/liter	
KNO_3	2500	1300
$MgSO_4 \cdot 7H_2O$	190	190
$(NH_4)_2SO_4$	270	135
$CaCl_2 \cdot 2H_2O$	150	150
$NaH_2PO_4 \cdot H_2O$	150	150
$MnSO_4 \cdot H_2O$	20	20
$ZnSO_4 \cdot 7H_2O$	2.0	2.0
$CuSO_4 \cdot 5H_2O$	0.05	0.05
KI	0.75	0.75
$CoCl_2 \cdot 6H_2O$	0.03	0.03
H_3BO_3	4.5	4.5
$Na_2MoO_4 \cdot 2H_2O$	0.06	0.06
$FeSO_4 \cdot 7H_2O$	28.0	28.0
Na_2EDTA	37.2	37.2
Pyridoxine HCl	2.0	2.0
Inositol	25.0	25.0
IBA	0.01	1.0
BA	6.0	0.01
Gelrite	2.0	2.0

ROSE, *Rosa* spp. Rosaceae.

Explant: 1–3 cm actively growing shoot tip.

Treatment: Remove all leaves except tiny terminal ones. Wash shoots in water with few drops of detergent. Mix in 1/10 bleach for 20 minutes. Rinse in 3 changes of sterile, distilled water for 2 minutes each.

Medium: Lay shorter explants on modified WPM agar medium. Place longer explants in modified WPM liquid medium with the lower half of the stem in the liquid in the bottom of a test tube. As the terminal and lateral shoots grow out, transfer 2–3 cm pieces to agar medium. Add BA at 1 mg/L for multiplication.

Light: 100–300 f.c. from cool white fluorescent light with 16 hrs light/8 hrs dark.

Temperature: 25°C.

Discussion: Shoots are best rooted directly from Stage II. Root initials have been aided by placing shoots in the dark. Place in potting mix under high humidity. Tissue culture is a viable alternative to grafting roses.

References: #34, #67, #68, #87.

ROSE MEDIUM

Compound	Stage I & II
	mg/liter
NH_4NO_3	400
$Ca(NO_3)_2 \cdot 4H_2O$	556
KH_2PO_4	170
$CaCl_2 \cdot H_2O$	96
$MgSO_4 \cdot 7H_2O$	370
K_2SO_4	990
$FeSO_4 \cdot 7H_2O$	27.8
Na_2EDTA	32.8
Glycine	2.0
Inositol	100
Pyridoxine HCl	0.5
BA	1.0
Thiamine HCl	1.0
Nicotinic acid	0.5
Sucrose	20000
Agar	6000
pH 5.7	

BLACKBERRY, *Rubus* spp. Rosaceae.

Explant: 5 cm shoot tips. Remove leaves except for youngest ones enclosing the terminal bud.

Treatment: Agitate shoot tips in distilled water with few drops of detergent for one minute. Rinse for 5 minutes each in four changes of distilled water. Mix for 5 minutes in 1/20 bleach with 0.1% Tween 20. Rinse briefly in sterile water followed by 5 minutes each in 3 changes of sterile water. Trim to 2–4 cm.

Medium: Modified MS with IBA, BA, and GA_3. Stage I may be in agitated liquid or on agar. Use Stage III or root directly from Stage II into fine greenhouse potting mix, milled sphagnum, or Jiffy 7's.

Light: 200–400 f.c. from warm white fluorescent lamps with 16 hrs light/8 hrs dark.

Temperature: 25°C.

Discussion: As a new shoot is produced by an explant, subculture the shoot with an attached piece of the original explant until the shoot reaches 2–5 cm. Blackberries respond well to culture with 3× to 6× multiplication every 3–4 weeks.

References: #11, #22, #40, #81, #113, #176.

BLACKBERRY MEDIUM

Compound	Stage I & II	III
	mg/liter	
MS salts	4628	4628
Inositol	100	100
Thiamine HCl	0.4	0.4
BA	1.0	—
IBA	—	1.0
GA_3	0.5	—
Sucrose	30000	30000
Gelrite	3000	3000
Agar	1000	1000
Charcoal	—	0.6
pH 5.2		

RASPBERRY, *Rubus spp.* Rosaceae

Explant: 2–3 inches of active new growth. Trim to single node sections or excise buds.

Treatment: Wash shoots in water with 0.1% Tween 20 for 15 minutes. Stir in 10% bleach for 10 minutes. Treatment time should correspond to tenderness of shoot. Rinse four times in sterile distilled water.

Media: This multiplication medium is the same as for blackberry. For in vitro rooting a modified Anderson medium is recommended.

Light: 100–300 f.c. from cool white fluorescent lights with 16 hrs light/8 hrs dark.

Temperature: 25°C.

Discussion: Most raspberries are difficult. The larger the explant the sooner it will start growing but the more apt it is to be contaminated. Excised buds will take longer but are more apt to be clean. In any case they usually take several weeks to react. Even so, transferring should take place every 2 weeks or more frequently. Experiment and fine tuning with hormone strength, gel strength, and salt concentrations is important to success.

In light of new cultivars being released regularly it is important to learn more about raspberry tissue culture so as to bring more of these new plants into the market place sooner.

Shade and humidity are critical in growing-on these tissue cultured plantlets. Jiffy 7's or light seedling mix will help the transition period during which they must grow new leaves before they will flourish in normal greenhouse environment.

References: #22, #40, #51, #81, #113, #176.

RASPBERRY MEDIA

Compound	Stage I & II	Stage III
	mg/liter	
MS salts	4628	—
Inositol	100	100
Thiamine HCl	0.4	0.4
BA	1.0	—
GA_3	0.5	—
NH_4NO_3	—	400
KNO_3	—	480
$NaH_2PO_4 \cdot H_2O$	—	380
$CaCl_2 \cdot 2H_2O$	—	440
$MgSO_4 \cdot 7H_2O$	—	370
MS minor salts	—	32.73
MS Iron	—	65.1
IBA	—	1.0
Sucrose	30000	30000
Charcoal	—	0.6
Gelrite	3000	3000
Agar	1000	1000
pH 5.3		

GRAPE, *Vitis vinifera.* Vitaceae.

Explant: 3 node shoot tips.

Treatment: Cut 5 cm shoot tips. Remove and discard the expanded leaves. Stir tips in 70% alcohol for 1 minute OR in 7% bleach plus 0.1 Tween for 20 minutes. Rinse 4 times in sterile distilled water. In the hood cut 3-node tips, and place on soft Gelrite and agar.

Media: Pint Mason jars are recommended in preference to smaller containers for Stage II. Stage I: ¾ MS salts with thiamine, BA, inositol, sucrose, Gelrite, and agar. Stage II: Add Na $H_2PO_4 \cdot H_2O$ and $AdSO_4$, increase BA, and increase Gelrite/agar. Stage III: Lower phosphate, inositol and sucrose. Omit $AdSO_4$ and BA. Add IAA at 0.1 mg/L, and soft Gelrite/agar. Or root directly in "soil" (Stage IV), but they may take longer to establish.

For this stage use 1 peat/4 vermiculite or coarse perlite. In either case water with Stage III medium without sucrose. Place in high humidity for 3 weeks followed by intermittent mist.

Light: 300 f.c. from fluorescent lights with 16 hrs light/8 hrs dark.

Temperature: 23°C. 30°C is preferred but not essential for rooting.

Discussion: The urgent need for virus-free, crowngall-free grapevines is beginning to be met. With 2 mm explants plants can be rendered virtually disease free for vineyards or stocks held in culture. When plants are disease free larger explants are advised for quicker establishment. By conventional methods it is extremely time consuming and costly to eradicate disease, and several more years are required to propagate a few hundred plants from a "clean" plant, or 2 years to produce field ready plants by grafting. In contrast, hundreds of tissue cultured plants can be field ready in a year, either on their own roots or by grafting of tissue cultured rootstock and scion.

Although prospects look bright for grape micropropagation, the genus remains difficult to manage due to inconsistent responses. Frequent transfer (2 week intervals), special concern for detail, and careful observation with prompt action when called for are particularly required for this important cultivar.

References: #18, #27, #28, #61, #79, #105, #113.

(Grape media next page)

GRAPE MEDIA

Compound	Stage I	Stage II (mg/liter)	Stage III
MS salts	3471 (= ¾ strength)	3471	3471
$NaH_2PO_4 \cdot H_2O$	—	170	150
Thiamine	0.4	0.4	0.4
$AdSO_4$	—	80	—
Inositol	100	100	25
BA	0.1	2.0	—
IAA	—	—	0.1
Sucrose	30000	30000	10000
Gelrite	1000	2000	1000
Agar	500	500	500
pH 5.3			

LEUCAENA LEUCOCEPHALA, K67. Leguminosae.

Explant: Single nodes from vegetative shoots from year old (plus) trees.

Treatment: Dip 3 cm shoot sections in 90% ethyl alcohol followed by stirring for 5 minutes in 20% bleach. Rinse in sterile water. Stir for 5 minutes in 0.1% mercuric chloride and rinse 3 times in sterile water.

Media: MS (or ½ MS) with inositol and BA for shoot growth. ½ MS with kinetin and IBA initiate rooting. A final transfer into White's liquid medium with bridges helps to condition roots.

Light: 100–300 f.c. from cool white fluorescent tubes, 16 hrs light/8 hrs dark.

Temperature: 28°C.

Discussion: Leguminosae have lagged behind many other families in tissue culture production by way of shoot culture. Surprisingly, they are comparatively slow and difficult.

Leucaena are prized for fuel, forage, biomass, and as ornamentals. New improved hybrids should respond to these procedures thus hasten their distribution.

To grow-on: Plant in peat/vermiculite/perlite with slow-release nutrients. Cover with polyethylene for 20 days but uncover for 1–2 hours a day. Do not cover after about 3 weeks by which time there should be new growth.

References: #54, #113.

LEUCAENA MEDIA

Compound	Stage I & II	Stage III (mg/liter)	Conditioning
MS salts	4628 or 2314 (= ½)	2314	—
Inositol	100	100	—
BA	3.0	—	—
IBA	—	3.0	—
Kinetin	—	0.5	—
KNO_3	—	—	80
$MgSO_4$	—	—	720
$Ca(NO_3)_2 \cdot 4H_2O$	—	—	300
Na_2SO_4	—	—	200
KCl	—	—	65
$NaH_2PO_4 \cdot H_2O$	—	—	16.5
H_3BO_3	—	—	1.5
$MnSO_4 \cdot H_2O$	—	—	5.3
$ZnSO_4 \cdot 7H_2O$	—	—	3.0
KI	—	—	0.75
$Fe_2(SO_4)_3$	—	—	2.5
Sucrose	30000	15000	15000
Agar	8000	8000	—
pH 5.8			

BEGONIA REX, Rieger Elatior. Begoniaceae.

Explant: Leaf petiole segments, 5 mm long.

Treatment: Rinse 10 minutes in 1/5 bleach. Rinse 3 times in sterile water.

Media: MS salts, vitamins, inositol, BA (0.4 mg/L for shoots, none for rooting), and NAA (0.1 mg/L for both shoots and roots). Reduce sugar to 20 mg/L for rooting.

Light: Use 100 f.c. from cool white fluorescent lights for Stages I and II. Use 16 hrs light/8 hrs dark for all stages. For Stage III use 300 f.c.

Temperature: 25°C.

Discussion: Rooting can also be effected by planting directly from multiplication to potting mix under high humidity.

References: #103, #113.

BEGONIA MEDIA

	Stage I & II	III
Compound	*mg/liter*	
MS salts	4628	4628
Inositol	100	100
Thiamine	1.5	1.5
Nicotinic acid	0.5	0.5
Pyridoxine HCl	0.5	0.5
Glycine	2.0	2.0
NAA	0.1	0.1
BA	0.4	—
Sucrose	30000	0
Agar	8000	8000
pH 5.5		

CACTUS, *Mammillaria elongata*. Cactaceae.

Explant: Tubercles with spines removed.

Treatment: Mix excised branches for 1 hour in a saturated solution of benomyl with 0.5% Tween 20 added. Trim off spines. Mix branches in 1/10 bleach for 30 minutes. Remove the tubercles and rinse in two rinses of sterile distilled water for two minutes each.

Media: MS salts with organics which include high sugar (45 g/L), L-tyrosine (100 mg/L), vitamins, inositol, and adenine sulfate. Growth regulators are 2iP (10 mg/L), IBA (1 mg/L), and IAA (0.5 mg/L).

Light: 100–300 f.c. from cool white fluorescent light with 16 hrs light/8 hrs dark.

Temperature: 27°C.

Discussion: For root initiation transfer to the greenhouse in pots containing 2 parts peat/1 part vermiculite/1 part perlite. Cover with plastic. To harden off, over a 2 week period gradually enlarge a pinhole size hole in the plastic cover.

References: #75, #91, #113.

MAMMILLARIA MEDIUM

	Stage I & II
Compound	*mg/liter*
MS salts	4628
$NaH_2PO_4 \cdot H_2O$	85
Inositol	100
Thiamine HCl	30
Adenine sulfate	80
L-Tyrosine	100
Pyridoxine HCl	1.0
Nicotinic acid	10
IAA	0.5
IBA	1.0
2iP	10
Sucrose	45000
Agar	8000
pH 5.5	

CACTUS, *Epiphyllum chrysocardium.* Cactaceae.

Explant: Stem cuttings 1.5–2 cm long.

Treatment: Dip in 30% alcohol. Rinse in sterile distilled water. Mix in 1/10 bleach for 15 minutes. Rinse 3 times in sterile distilled water.

Media: Half strength MS salts with inositol (500 mg/L), pyridoxine (5 mg/L), thiamine (1 mg/L), and sucrose (20 g/L). Recommended hormones are BA (1 mg/L) and NAA (0.1 mg/L). For rooting omit BA and NAA and add 0.01 mg/L of IBA.

Light: Sylvania Gro-Lux lamps were used with 16 hrs light/8 hrs dark.

Temperature: 26°C.

Discussion: This attractive tropical cactus is known as Heart of Gold because of its large white flowers and yellow filaments. The sunken shoot primordia can be expected to emerge in two weeks. Shoots should have 5 or more buds before being transferred to rooting medium.

References: #84, #91, #113.

EPIPHYLLUM MEDIA

	Stage I & II	III
Compound	*mg/liter*	
MS salts	2314	2314
Inositol	500	500
Pyridoxine HCl	5.0	5.0
Thiamine HCl	1.0	1.0
BA	1.0	—
NAA	0.1	—
IBA	—	0.01
Sucrose	20000	20000
Agar	8000	8000
pH 5.7		

EUCALYPTUS FICIFOLIA. Myrtaceae.

Explant: Nodes and apical buds from seedlings, adult trees, or coppice (coppice is a juvenile material growing from the base of adult trees, sometimes in response to injury).

Treatment: Cut off leaves leaving about 1 cm of petiole. Wash nodes in running tap water for 1 hour. Wash in distilled water with 0.1% Tween 20 for 5 minutes. Agitate in 5% calcium hypochlorite (freshly prepared and filtered) for 20 minutes. Rinse briefly in three sterile distilled water rinses. Soak in ascorbic acid solution (2 mg/L) for 2 hours. Rinse in sterile, distilled water. Trim to about 1 cm of main stem at node or apex. Place on agar medium and incubate for 24 hours in the dark.

Media: One half strength MS with low sugar (20000 mg/L), BA and NAA for multiplication, and changing to IBA for rooting is a simple alternative among various formulas published.

Light: Place tubed explant in dark for 24 hrs. Move to 100–300 f.c. from cool white fluorescent lights with 16 hrs light/8 hrs dark.

Temperature: 25°C.

Discussion: Multiple buds may be expected to appear in two months from initial explant. During this period there should be several transfers to fresh medium. Roots may appear in three weeks after being placed in rooting medium. Other *Eucalyptus* species have been cultured with varied success on the same or different media, but *E. filicifolia* appears to be the most successful. Several major problems are (1) Contamination, particularly with adult material; (2) "Bleeding" (phenolic exudates), which usually occurs in freshly transferred material; this problem is minimized by the use of ascorbic acid soaking and initial incubation in the dark; (3) Rooting and establishment: Rooting percentage can be very low especially with certain cultivars; establishment in soil (1 sand/1 loam) takes a great deal of care. Humidity is lowered very gradually. The plants are covered with plastic wrap and placed in deep shade for 24 hours. They are then placed in 80% shade. After growth has started lighting is increased very gradually up to full sun. Leaves are subject to scorching with abrupt change to brighter light intensity.

References: #17, #36, #37, #65, #66, #113.

EUCALYPTUS MEDIA

Compound	Stage I & II	III
	mg/L	
MS salts	2314 (= ½ strength)	2314
BA	0.2	—
NAA	0.16	—
IBA	—	2.0
Sucrose	20000	20000
Agar	6000	8000
pH 5.5		

KALMIA LATIFOLIA. Ericaceae.

Explant: 2–3 cm shoot tips.

Treatment: Carefully remove all leaves over 1 cm. Dip in 70% ethyl alcohol. Mix in 1/10 bleach with 0.1% Tween 20 for 15 minutes. Rinse in 3 rinses of sterile distilled water for one minute each. Trim away any damaged material and place in liquid medium.

Media: Liquid woody plant medium (WPM) is a significant change from MS with low nitrate, high potassium, and low sugar. The minor elements, vitamins, and hormones remain fairly standard with 1 mg/L 2iP used for multiplication. Place the explants in agitated liquid WPM for one week. Change the medium daily. After one week transfer to stationary tubes with the explant bases in liquid medium. After 2 months place the axillary shoots on agar medium. If "bleeding" occurs transfer to fresh medium immediately.

Light: Continuous light from cool white fluorescent lights with 100 to 300 f.c.

Temperature: 28°C–30°C.

Discussion: Four week multiplication rate has been reported at 8X. Stage III is optional and plantlets may be rooted directly in 100% peat, 30°C, high humidity, and continuous light. *Kalmia* can also be cultured by following the directions and media for *Rhodod endrons.*

References: #13, #14, #87, #113.

KALMIA MEDIA

Compound	Stage I & II	III
	mg/liter	
NH_4NO_3	400	400
$CaCl_2 \cdot 2H_2O$	96	96
$MgSO_4 \cdot 7H_2O$	370	370
KH_2PO_4	170	170
$Ca(NO_3)_2 \cdot 4H_2O$	556	556
K_2SO_4	990	990
$FeSO_4 \cdot 7H_2O$	27.8	27.8
Na_2EDTA	37.3	37.3
MS minor salts	32.8	32.8
Inositol	100	100
Nicotinic acid	0.5	0.5
Pyridoxine	0.5	0.5
Thiamine HCl	1.0	0.1
Glycine	2.0	2.0
2iP	1.0	—
Sucrose	20000	20000
Agar	6000	6000
pH 5.2		

RHODODENDRON spp. Ericaceae.

Explant: 2–4 cm shoot tips of actively growing plants with lateral buds showing.

Treatment: Wash 5 cm shoot tips in water with few drops of detergent. Remove leaves and terminal bud. Dip in 70% alcohol for 10 seconds. Rinse briefly in distilled water. Mix in 1/10 bleach with 0.1% Tween 20 for 20 minutes. Rinse for 1 minute in 3 rinses of sterile distilled water. Trim away 1 cm from base. Lay tips firmly on agar slant, making sure of good contact between shoot and agar.

Media: Use Anderson's rhododendron formula, a modified MS with lowered nitrate, increased iron, sodium phosphate, adenine sulfate, 2iP (5 mg/L), and IAA (1 mg/L). For Stage III use ⅓ strength salts, omit 2iP and IAA, add 5 mg/L of IBA and .8 g/L charcoal, or omit Stage III and root directly into 1 peat/1 perlite under mist.

Light: 100–300 f.c. from cool white fluorescent light with 16 hrs light/8 hrs dark.

Temperature: 25°C.

Discussion: Transfer new explants every 2 weeks until growth appears, then every 4–6 weeks. As the lateral buds break, the explant stem is gradually cut away in 2 or 3 transfers. By the time the leaves are completely miniaturized, 3 to 4-fold multiplication can be expected every 6 weeks.

(See Figure 12).

References: #6, #13, #14, #27, #82, #93, #100, #137, #169.

RHODODENDRON MEDIA

Compound	Stage I & II	III
	mg/liter	
NH_4NO_3	400	133
KNO_3	480	160
$NaH_2PO_4 \cdot H_2O$	380	127
$CaCl_2 \cdot 2H_2O$	440	147
$MgSO_4 \cdot 7H_2O$	370	123
$FeSO_4 \cdot 7H_2O$	55.7	18.5
Na_2EDTA	74.5	24.8
H_3BO_3	6.2	2.0
$MnSO_4 \cdot H_2O$	16.9	5.6
$ZnSO_4 \cdot 7H_2O$	8.6	2.9
$Na_2MoO_4 \cdot 2H_2O$	0.25	0.08
$CuSO_4 \cdot 5H_2O$	0.025	0.008
$CoCl_2 \cdot 6H_2O$	0.025	0.008
Inositol	100	100
Thiamine HCl	0.4	0.4
IAA	1.0	—
2iP	5.0–15.0	—
IBA	—	5.0
Adenine sulfate	80	—
Sucrose	30000	30000
Charcoal	—	800
Agar	6000	6000
pH 4.5–5.0		

BLUEBERRY, *Vaccinium* spp. Ericaceae.

Explant: 1–4 cm shoot tips from actively growing shoots.

Treatment: Cut 2–5 cm actively growing shoot tips and stem pieces with lateral buds. Trim off leaves without damaging axillary buds. Wash in distilled water with few drops of detergent. Rinse in sterile, distilled water with 0.1% Tween 20 for 5 minutes. Place in freshly prepared calcium hypochlorite solution prepared as follows: Add 60 grams calcium hypochlorite to one liter of distilled water. Stir for 15 minutes. Allow to settle and decant solution. Mix solution with equal parts of distilled water and 0.1% Tween 20. Stir explants in calcium hypochlorite solution for 10 minutes. Rinse for 10 seconds in each of 2 rinses of sterile, distilled water. Dip in sterile, antioxidant solution (citric acid/ascorbic acid 1/1). Rinse for 2 minutes in sterile, distilled water. Trim shoot tips to 1–4 cm. Trim axillary buds to have 5 mm adjoining stem either side of node. Lay explants on agar and press lightly into the agar so that good contact is made between total length of stem and the agar.

Media: The macronutrients are high in calcium nitrate and potassium phosphate, and without calcium chloride. MS micronutrients used are the same as MS, but the iron is double the usual amount. Use 2iP (15 mg/L) and IAA (4 mg/L). In the last transfer prior to rooting omit 2iP to permit greater shoot elongation. Older cultures may benefit from lowering the hormones, or leaving them out altogether for one or more transfer cycles.

Light: 200–400 f.c. from warm white fluorescent lights with 16 hrs light/8 hrs dark.

Temperature: 24°C–26°C.

Discussion: An alternate medium is that used for *Rhododendron.* For rooting, remove from culture shoots that are at least 2–3 cm long. Dip the shoot bases into IBA solution prepared by mixing 100 mg IBA in 100 ml of 50% ethyl alcohol. Plant in milled sphagnum in high humidity. Roots start in 3–7 weeks.

References: #13, #14, #113, #115, #177.

BLUEBERRY MEDIA

Compound	Stage I & II	III or omit
	mg/liter	
NH_4NO_3	160	160
KNO_3	202	202
$MgSO_4 \cdot 7H_2O$	370	370
KH_2PO_4	408	408
$Ca(NO_3)_2 \cdot 4H_2O$	708	708
$FeSO_4 \cdot 7H_2O$	55.7	55.7
Na_2EDTA	74.4	74.4
MS minor salts	32.8	32.8
Inositol	100	100
Thiamine HCl	0.4	0.4
Adenine sulfate	80	—
IAA	4.0	—
IBA	—	5.0
2iP	15	—
Sucrose	30000	30000
Agar	5500	5500
pH 4.8		

PRIMULA ACAULIS. Primulaceae.

Explant: Remove small division from parent plant. Wash in tap water with few drops of liquid detergent. Cut away and discard outer leaves. Carefully excise shoot bud with a few millimeters of base.

Treatment: Agitate excised shoot bud in 1/10 bleach with 0.1% Tween 20 for 20 minutes. Rinse in 1/100 bleach followed by 4 or 5 washes in sterile distilled water.

Media: Anderson's inorganics and organics especially formulated for double acaulis.

Light: 100–300 f.c. from cool white fluorescent lamps. 16 hrs light/8 hrs dark.

Temperature: 25°C.

Discussion: Wilbur Anderson, who pioneered rhododendron culture, has again researched a new field. He developed this formula especially for double *P. acaulis* to increase clonal lines for F1 hybrid seed production. Primroses are difficult to clean because the shoot tip area is in such close contact with the soil.

References: #7, #132.

PRIMULA MEDIA

Compound	Stage I & II	Stage III
	mg/liter	
KNO_3	480	240
NH_4NO_3	400	200
$CaCl_2 \cdot 2H_2O$	440	220
$MgSO_4 \cdot 7H_2O$	370	185
$NaH_2PO_4 \cdot H_2O$	380	190
H_3BO_3	6.2	3.1
$MnSO_4 \cdot H_2O$	16.9	8.5
$ZnSO_4 \cdot 7H_2O$	8.6	4.3

(continued next page)

(Primula media continued)

Compound	Stage I & II	Stage III
	mg/liter	
KI	0.30	0.15
$Na_2MoO_4 \cdot 2H_2O$	0.25	0.13
$CuSO_4 \cdot 5H_2O$	0.025	0.013
$CoCl_2 \cdot 6H_2O$	0.025	0.013
$FeSO_4 \cdot 7H_2O$	55.7	27.9
$Na_2EDTA \cdot 2H_2O$	74.5	37.3
Inositol	100	100
$AdSO_4$	80	—
Thiamine HCl	0.4	0.4
IAA	1.5	0.5
BA	2.5	—
Sucrose	30000	30000
Agar	6000	6000
pH 5.7		

PHLOX SUBULATA. Polemoniaceae.

Explant: 0.5–1 mm shoot tips excised from actively growing, cleaned shoots.

Treatment: Remove leaves over 1 cm long from 2 cm long shoots. Mix shoots in 1/100 bleach for 25 minutes. Rinse in sterile, distilled water by dipping briefly into 10 separate rinses. Excise meristem with 1 or 2 pairs of leaf primordia.

Media: Modified MS salts with Nitsch additives (biotin, glycine, pyridoxine HCl, thiamine HCl, nicotinic acid, and folic acid), GA_3 (.0035 mg/L), and BA (5 mg/L). Rooting medium is the same except GA_3 and BA are omitted and NAA is added (0.5 mg/L).

Light: 100 f.c. from Gro-Lux lamps with 16 hrs light/8 hrs dark.

Temperature: 22°C–26°C.

Discussion: Tissue culture of *Phlox* can assist in eliminating disease transmitted by the usual crown division method of multiplication. Crown division also has the disadvantage of being limited to periods of dormancy; multiplication by tissue culture is not limited by season.

References: #113, #132.

PHLOX MEDIA

Compound	Stage I & II	III
	mg/liter	
MS salts	4628	4628
Inositol	100	100
Adenine sulfate	40	—
Thiamine HCl	0.5	0.5
Nicotinic acid	0.5	0.5
Pyridoxine HCl	0.5	0.5
Folic acid	0.4	—
Glycine	2.0	2.0
Biotin	—	0.2
GA_3	.004	—
NAA	—	0.5
BA	5.0	—
Sucrose	30000	30000
Agar	6000	6000
pH 5.5		

POTATO, *Solanum tuberosum*. Solanaceae.

Explant: Apical and axillary buds from sprouted tubers.

Treatment: Dip shoots in 70% alcohol. Stir 10 minutes in 1/10 bleach plus 0.1% Tween 20. Rinse 3 times in sterile distilled water.

Medium: MS with inositol, thiamine, kinetin, and GA_3.

Light: 16 hrs light/8 hrs dark with 300 f.c. from cool white fluorescent lights.

Temperature: 23°C.

Discussion: Because potatoes are very subject to disease, tissue culture is an ideal way to multiply disease-free stock. The method of micropropagation is nodal propagation. Establishment of the primary explant requires approximately 8 weeks. Nodal multiplication rates vary from variety to variety and from line to line (a line is all those plants derived from a single tuber). On this particular medium, roots develop along with shoots. A separate rooting medium is not needed. Transfer to soil is readily accomplished in a mist bench or a humidity tent.

References: #52, #107, #113, #155.

POTATO MEDIUM

Compound	Stage I, II, & III *mg/liter*
MS salts	4628
Inositol	100
Thiamine HCl	0.4
GA_3	0.1
Kinetin	0.5
Sucrose	30000
Agar	7000
pH 5.7	

AFRICAN VIOLET, *Saintpaulia ionantha* Wendle. Gesneriaceae.

Explant: 1 cm square leaf pieces and 2 mm cross sections of petiole.

Treatment: Wash whole leaf and petiole in water with 1% detergent. Agitate in 1/10 bleach with 0.1% Tween 20 for 10 minutes. Rinse in three rinses of sterile, distilled water. Cut leaves into 1 cm squares and petioles into 2 mm cross sections.

Media: MS salts with 0.1 mg/L of NAA and 0.1 mg/L of BA for petiole cross sections. For leaf sections use a modified MS with 2 mg/L IAA and .08 mg/L of BA.

Light: 300 f.c. from cool white fluorescent lamps. 16 hrs light/8 hrs dark. Stage III light intensity for rooting: 50–100 f.c.

Temperature: 25°C.

Discussion: It is not surprising that African violets are easily tissue cultured in view of the fact that they usually propagate so readily from leaf cuttings. The decision to culture African violets *in vitro* is a matter of economics which must be made by the individual grower.

References: #31, #113.

AFRICAN VIOLET MEDIA

	Stage I, II & III	
	Leaf sec.	Petiole × sec.
Compound	*mg/liter*	
MS salts	4628	4628
NaH_2PO_4	170	170
Inositol	100	—
Pyridoxine	0.4	—
Niacin	0.4	—
Thiamine	0.4	—
BA	.08	.01
IAA	2.0	—
NAA	—	0.1
Sucrose	30000	30000
Agar	8000	8000
pH 5.5		

CUCUMBER, *Cucumis sativus* L. (gynoecious cucumber). Cucurbitaceae.

Explant: Axillary buds 1–3 mm long taken from month-old plants.

Treatment: Place leaf axils in 1/10 bleach for 5 minutes. Rinse in 3 changes of sterile, distilled water for 2 minutes each. Excise 1–3 mm long axillary buds.

Media: MS salts plus vitamins, NAA and BA.

Light: Cool white fluorescent tubes and incandescent bulbs providing 400 f.c. with 16 hrs light/8 hrs dark.

Temperature: 25°C.

Discussion: Gynoecious hybrids produce all, or primarily, female flowers. Tissue culture offers a convenient method for multiplying these plants as opposed to cuttings, which are usually scarce, or chemically inducing staminate flowers.

References: #58, #113, #161.

CUCUMBER MEDIA

	Stage I & II	III
Compound	*mg/liter*	
MS salts	4628	4628
Inositol	100	100
Nicotinic acid	0.5	0.5
Pyridoxine HCl	1.0	—
Thiamine HCl	1.0	—
BA	1.0	—
NAA	0.1	—
Agar	7000	—
Sucrose	30000	—
pH 5.5		

GERBERA JAMESONII Bolus. Compositae.

Explant: 3–10 mm shoot tip.

Treatment: Use as clean starting material as possible. Obtain emerging shoot tips together with 2–3 mm of base from the crown. Wash briefly in running water. Mix in three changes of water with 1.0% Tween 20 for 10 minutes. Rinse in three sterile water rinses for 2 minutes each. Soak for 5 minutes in 1/20 bleach. Under the microscope excise emerging buds with 1 mm of base. Dip for 1 minute in 1/20 bleach and place on agar medium.

Media: Modified MS salts with 0.5 mg/L IAA and 10 mg/L kinetin for Stage I and II. Stage III has no kinetin or adenine sulfate but increase IAA to 10 mg/L. Two weeks in Stage III medium is sufficient before transplanting to potting mix.

Light: 100 f.c. from GroLux lights with 16 hrs light/10 hrs dark. For Stage III rooting use 1000 f.c. from cool white fluorescent lights.

Temperature: 27°C.

Discussion: Although gerberas grow easily from seed, they are often tissue cultured to produce large numbers of specified colors and to multiply desirable individuals. Contamination is a persistent problem. Tissue culture of seedlings from sterilized seed of known parentage should be considered as an alternative.

References: #112, #113, #124.

GERBERA MEDIA

	Stage I & II	III
Compound	*mg/liter*	
MS salts	4628	4628
NH_2PO_4	85	85
Nicotinic acid	10	10
Pyridoxine	1.0	1.0
Thiamine	30	30
Adenine sulfate	80	—
IAA	0.5	10
Kinetin	10	—
Sucrose	45000	45000
Agar	8000	8000
pH 5.5		

PYRETHRUM, *Chrysanthemum cinerariaefolium*. Compositae.

Explant: Young flower heads.

Treatment: Wash young, unopened flower heads in water with few drops of detergent. Rinse twice in distilled water. Mix in 5% calcium hypochlorite solution for 30 minutes. Rinse for 30 minutes in sterile, distilled water. Cut the heads in two longitudinally. Remove and discard bracts, disc flowers and ray flowers.

Medium: A combination of half strength Knop's macronutrients and half strength Heller's micronutrients is combined with BA (1 mg/L).

Light: 300 f.c. from cool white fluorescent lights with 14 hrs light/8 hrs dark.

Temperature: 18°C during light, 14°C during dark.

Discussion: Shoots develop in 3 weeks. Most are very short. Dip 1 cm long shoots in IAA (1% in talc) and root in potting mix. Roots form in about 3 weeks. Other composites that have responded to above procedures are: *C. parthenium, C. leucanthemum, C. segetum, Anthemis arvensis, Calendula officinalis, Hypochaeris radicata, H. autumnale, Leontodon autumnalis,* and *Matricaria maritima.*

References: #85, #128.

PYRETHRUM MEDIUM

Compound	Stage I & II *mg/liter*
NH_4NO_3	62.5
$Ca(NO_3)_2 \cdot 4H_2O$	250
KH_2PO_4	62.5
$MgSO_4 \cdot 7H_2O$	62.5
$FeSO_4 \cdot 7H_2O$	.56
Na_2EDTA	.75
$MnSO_4 \cdot H_2O$	.05
$ZnSO_4 \cdot 7H_2O$	0.5
H_3BO_3	0.5
KI	.005
$CuSO_4 \cdot 5H_2O$	.015
BA	1.0
Sucrose	50000
Agar	6000
pH 5.7	

Appendices

SOME BASICS OF METRICS

Because some lab equipment and media formulas are calibrated or expressed in metric terms, it is important to be able to read and understand them. Familiarity with metric measurements develops rapidly with practice and makes media calculations relatively easy. It is convenient to remember a few metric facts; for example, the fact that one milligram in a liter of solution equals one part per million (ppm).

Facts to remember:

Weight—
- .001 gram (g) = 1 milligram (mg)
- 1000 mg = 1 g
- 1000 g = 1 kilogram (kg)
- 1 kg = 2.2 pounds
- 1 pound = 454 g

Volume—
- 1000 milliliters (ml) = 1 liter (l or L)

Length—
- 10 millimeters (mm) = 1 centimeter (cm)
- 2.54 cm = 1 inch

Temperature—
- 77°F = 25°C
- To convert Fahrenheit (F) to Celcius (C):
  - $C = 5/9\ (F - 32)$
- To convert Celcius to Fahrenheit:
  - $F = 9/5\ (C + 32)$

Concentration—
- 1 mg/l = 1 ppm (parts per million)
- 1 molar solution = one mole in a L of solution
- 1 mole = the molecular weight of a chemical expressed in grams

FORMULA COMPARISON CHART

mg/liter

	Anderson (1978)	Gamborg (1968)	Gautheret (1942)	Heller (1953)	Hildebrandt, Riker & Dugan (1946)	Hoagland (1950)	Knop (1865)	Knudson C (1946)	Linsmaier & Skoog (1965)	McCown & Lloyd (1980) (WPM)	Morel & Muller (1964)	Murashige & Skoog (1962)	Nitsch & Nitsch (1969)	Schenk & Hildebrandt (1972)	Vacin & Went (1949)	White (1963)
NH_4NO_3	400	—	—	—	—	—	—	—	1650	400	—	1650	720	—	—	—
KNO_3	480	2500	125	—	160	607	125–200	—	1900	—	—	1900	950	2500	525	80
$CaCl_2 \cdot 2H_2O$	440	150	—	75	—	—	—	—	400	96	—	440	166	200	—	—
$MgSO_4 \cdot 7H_2O$	370	250	125	250	720	250	125–200	250	370	370	125	370	185	400	250	720
KH_2PO_4	—	—	125	—	—	—	125–200	250	170	170	125	170	68	—	250	—
$(NH_4)_2SO_4$	—	134	—	125	—	—	—	500	—	—	1000	—	—	—	500	—
$NaH_2PO_4 \cdot H_2O$	380	150	125	—	132	—	—	—	—	—	—	—	—	—	—	16.5
$Ca_3(PO_4)_2$	—	—	—	—	—	—	—	—	—	—	—	—	—	—	200	—
$Ca(NO_3)_2 \cdot 4H_2O$	—	—	500	—	800	945	500–800	1000	—	556	500	—	—	—	—	300
Na_2SO_4	—	—	—	—	100	—	—	—	—	—	—	—	—	—	—	200
KCl	—	—	—	750	130	—	—	—	—	—	1000	—	—	—	—	65
K_2SO_4	—	—	—	—	—	—	—	—	—	990	—	—	—	—	—	—
$NaNO_3$	—	—	600	600	—	—	—	—	—	—	—	—	—	—	—	—
$NH_4H_2PO_4$	—	—	—	—	—	115	—	—	—	—	—	—	—	300	—	—
H_3BO_3	6.2	3.0	0.05	1.0	3.0	2.86	—	—	6.2	6.2	—	6.2	10	5.0	—	1.5
$MnCl_2 \cdot 4H_2O$	—	—	—	—	—	1.81	—	—	—	—	—	—	—	—	—	7.0
$MnSO_4 \cdot H_2O$	16.9	10	3.0	—	—	—	—	—	—	22.3	—	16.9	—	10	—	—
$MnSO_4 \cdot 4H_2O$	—	—	—	0.1	4.5	—	—	7.5	22.3	—	—	—	25	—	7.5	—
$ZnSO_4 \cdot 7H_2O$	8.6	2.0	0.18	1.0	3.0	—	—	—	8.6	8.6	—	8.6	10	1.0	—	3.0
KI	—	0.75	0.5	0.1	0.375	—	—	—	0.33	—	—	0.83	—	1.0	—	0.75
$Na_2MoO_4 \cdot 2H_2O$	0.25	0.25	—	—	—	—	—	—	0.25	0.025	—	0.25	0.25	0.1	—	—
$CuSO_4 \cdot 5H_2O$	0.025	0.025	0.05	—	—	—	—	—	0.025	.025	—	0.025	0.025	0.2	—	—
$CoCl_2 \cdot 6H_2O$	0.025	0.025	0.05	—	—	—	—	—	0.025	—	—	0.025	—	0.1	—	—
$FeSO_4 \cdot 7H_2O$	55.7	27.8	—	—	—	5	—	25	2.8	27.8	—	27.8	27.8	15	—	—
$Fe_2(SO_4)_3$	—	—	—	50	—	—	—	—	—	—	—	—	—	—	—	2.5
$Na_2EDTA(2H_2O)$	74.5	37.3	—	—	—	—	—	—	37.3	37.3	—	37.3	37.3	20	—	—
Fe tartrate	—	—	—	—	5.0	—	—	—	—	—	—	—	—	—	28	—
Biotin	—	—	—	—	—	—	—	—	—	—	—	—	0.05	—	—	—

	Anderson (1978)	Gamborg (1968)	Gautheret (1942)	Heller (1953)	Hildebrandt, Riker & Dugan (1946)	Hoagland (1950)	Knop (1865)	Knudson C (1946)	Linsmaier & Skoog (1965)	McCown & Lloyd (1980) (WPM)	Morel & Muller (1964)	Murashige & Skoog (1962)	Nitsch & Nitsch (1969)	Schenk & Hildebrandt (1972)	Vacin & Went (1949)	White (1963)
Inositol	100	100	—	—	—	—	—	—	100	100	—	100	100	1000	—	—
Adenine SO_4	80	—	—	—	—	—	—	—	—	—	—	—	—	—	—	—
Nicotinic acid	—	1.0	—	—	—	—	—	—	—	0.5	—	0.5	5.0	5.0	—	0.5
Thiamine · HCl	0.4	10	—	—	—	—	—	—	—	1.0	—	0.1	0.5	5.0	—	0.1
Pyridoxine · HCl	—	1.0	—	—	—	—	—	—	—	0.5	—	0.5	0.5	0.5	—	0.1
Glycine	—	—	—	—	—	—	—	—	—	2.0	—	—	2.0	—	—	3.0
Sucrose	30000	30000	—	—	—	—	—	—	30000	20000	—	30000	20000	—	—	20000
IAA	1.0	—	—	—	—	—	—	—	1.30	—	—	1–30	0.1	—	—	—
2iP	5.0	—	—	—	—	—	—	—	—	—	—	—	—	—	—	—
Kinetin	—	—	—	—	—	—	—	—	0.001–10	—	—	.04–10	—	—	—	—
Agar	6000	—	—	—	—	—	—	—	10000	6000	—	1000	8000	—	—	—
References	#13	#47	#49	#39	#72	#127	#127	#78	#86	#92	#109	#113	#116	#131	#156	#163

1. These are well known basic formulations to which are added micronutrients, vitamins, growth regulators, sucrose, etc., as desired.
2. Omitted are compounds of (a) beryllium, (a) titanium, (a,b) nickle, and (c) folic acid.

SUPPLIERS

Supplier	Products / Telephone
Aldrich Chemical Co. 940 West Saint Paul Ave. Milwaukee, WI 52333	Chemicals 800-558-9160
Biological Supply Co. 12402 Evergreen Dr. Lynwood, WA 98037	(Glassware, microscopes, etc.; small orders accepted) 206-743-4270
Cadillac Plastic and Chemical Co. 2427 6th S., Seattle, WA 98108	(Polycarbonate sheets, see page 36) 206-682-7252
Carolina Biological Supply Co. 2700 York Rd. Burlington, North Carolina 27215 or Box 187 Gladstone, OR 97027	(Media, supplies, hoods, HEPA filters) 800-334-5551 800-547-1733
Flanders Filters, Inc. P.O. Box 1219 Washington, North Carolina 27889	(Laminar air flow hoods, HEPA filters) 919-946-8081
Flow Laboratories, Inc. 936 W. Hyde Park Blvd. Inglewood, CA 90302 or 1710 Chapman Ave. Rockville, Maryland 20852	(Media) 213-674-2700 301-881-2900
Fungi Perfecti P.O. Box 7634 Olympia, WA 98507	(Sterile culture and growing room equipment and supplies, desktop laminar flow hoods) 206-786-1105
GIBCO (Grand Island Biological Co.) 519 Aldo Ave. Santa Clara, CA 95050 or 3175 Staley Rd. Grand Island, N.Y. 14072	(Media) 408-988-7611 800-828-6686 (N.Y. State: 800-462-2555)
W. W. Grainger, Inc. 2001 Grand St. Seattle, WA 98144	(Motors) 206-251-5030
Integrated Air Systems, Inc. 3750 Cohasset St. Burbank, CA 91504	(Laminar air flow hoods, HEPA filters) 213-842-5211

KC Biological, Inc. P.O. Box 5491 Lenexa, Kansas 66215	(Media) 800-255-6032
Kelco 8355 Aero Dr. San Diego, CA 92123	(Gelrite™, agar substitute) 714-292-4900
Lab Equipment Magazine P.O. Box 14000 Dover, NJ 07801-9873	(Equipment information, vendors; free subscription)
Magenta Corp., John Song 3800 N. Milwaukee Ave. Chicago, Ill. 60641	(Containers) 312-777-5050
Monarch Marking Box 1403 Dayton, OH	(Labelers) 1-800-543-6650
New Brunswick Scientific Co., Inc. 1130 Somerset St. New Brunswick, N.J.	(Rotators)
Northern Steel 5501 1st Ave. South Seattle, WA 98108	(Slotted angles for shelving supports) 206-767-5383 (Washington state: 800-562-1086)
Scientific Supply and Equipment, Inc. 1818 E. Madison Seattle, WA 98122	(Chemicals, supplies, equipment) 800-426-0455 (Washington state: 800-552-7164)
Harry Sharp & Son 420 8th Ave. North Seattle, WA 98108	(Todd Planter flats) 800-426-7766 (Washington state: 800-552-9454)
Sigma Chemical Co. P.O. Box 14508 St. Louis, Missouri 63178	(Chemicals) 800-325-3010
Simon Keller LTD Lyssachstrasse 83 3400 Burgdorf, Switzerland	Hot bead sterilizer for instruments
Quality Water, Bob Fuller 381 Circulo Loredo Rohnert Park, CA 94928	(Water stills) 707-585-7440
VWR Scientific P.O. Box 3551 Seattle, WA 98124	206-575-1500 201-756-8030

PROFESSIONAL ORGANIZATIONS

International Association for Plant Tissue Culture
Dr. Roberta H. Smith, National Correspondent (USA)
Department of Plant Sciences
Texas A&M University
College Station, TX 77843
Publications: *Newsletter* and *Proceedings of the 5th International Congress of Plant Tissue and Cell Culture.*

Tissue Culture Association, Inc.
1 Bank Street, Suite 210
Gaithersburg, MD 20878
Publications: *In Vitro* and *TCA Report. Plant Cell, Tissue and Organ Culture.*

American Society for Horticultural Science
701 N. St. Asaph St.
Alexandria, VA 22314
Publications: *Journal of the American Society for Horticultural Science. HortScience.*

The International Plant Propagators' Society, Inc.
P.O. Box 3131
Boulder, Colorado 80307
Publications: *The Plant Propagator* and *The International Plant Propagators' Society—Combined Proceedings* (Annual).

ATOMIC WEIGHT OF ELEMENTS FREQUENTLY USED IN TISSUE CULTURE

Element	*Symbol*	*Atomic Weight*
Boron	B	10.811
Calcium	Ca	40.08
Carbon	C	12.01115
Chlorine	Cl	35.453
Cobalt	Co	58.9332
Copper	Cu	63.54
Hydogen	H	1.00797
Iodine	I	126.9044
Iron	Fe	55.847
Magnesium	Mg	24.312
Manganese	Mn	54.938
Molybdenum	Mo	95.94
Nitrogen	N	14.0067
Oxygen	O	15.9994
Phosphorus	P	30.9738
Potassium	K	39.102
Sodium	Na	22.9898
Sulfur	S	32.064
Zinc	Zn	65.37

Glossary

aberrant: different appearance from the normal.

abscission: the normal separation of leaves or fruit from plants by means of a thin walled layer of cells.

adenine: a nitrogen containing compound in nuclear material, used in sulfate form in tissue culture media.

adventitious: growing from unusual locations such as aerial roots from stems, or buds at other places than leaf axils or stem tips.

agar: a polysaccharide (literally, many sugars) gel derived from certain algae.

alkaloid: any of a number of colorless, crystalline, bitter organic substances, such as caffeine, quinine, and strychnine, that have alkaline properties. Many are found in plants; they can have toxic effects on humans.

amino acids: a group of nitrogenous organic compounds that serve as structural units of proteins.

anther culture: tissue culture of anthers to obtain haploid clones.

apex: tip.

apical: (adj.) tip, or apex.

apical meristem: meristem located at the apices of main and lateral shoots.

apices: plural of apex.

aseptic: free from microorganisms.

asexual: without sex, vegetative.

atom: the smallest particle of an element that still retains the characteristics of that element.

atomic weights: a system of relative weights of atoms.

autoclave: a vessel for sterilizing with steam under pressure.

autotrophic: capable of making its own food from inorganic substances. See heterotrophic.

auxins: growth hormones associated with cell division and enlargement, and root initiation.

axenic: free from contaminants.

axil: the point of attachment of lateral growth with the stem.

balance: a type of scale for weighing.

ballast: a type of transformer wired into fluorescent lighting.

biosynthesis: formation of chemicals by living cells.

"bleeding": used here to describe the occasional purplish black coloration of media due to phenolic products given off by (usually fresh) transfers.

"bridge": a piece of filter paper or paper toweling placed within a test tube of liquid medium to hold the culture out of the liquid and serve as a wick.

callus: an unorganized, proliferating mass of mostly undifferentiated cells.

caneberry: any berry plant with a cane, such as raspberry or blackberry.

carbohydrates: organic compounds, including sugars, starches, and celluloses, composed of carbon, hydrogen and oxygen.

carcinogen: substance that causes cancer.

casein hydrolysate: an amino acid found in liquid endosperm, often used as a helpful supplement in tissue culture media.

catalyst: a substance that affects the rate of a reaction without otherwise being involved in that reaction.
cell: the basic physical unit of living organisms.
cell culture: the culture of single cells, often included in the broad term, tissue culture.
chimeras: plants with tissues of more than one genetic make up, such as some plants with variegated leaves.
chlorosis: an absence of green pigments in plants due to lack of light, or a magnesium or iron deficiency or other causes.
clone: the plants produced asexually from a single plant.
coconut milk: (coconut water): the liquid endosperm from coconuts.
coenzyme: a substance which activates an enzyme or accelerates its action.
compound: two or more elements chemically combined in fixed proportions.
contaminants: used here as all microorganisms.
cultivar: a named plant variety under cultivation.
culture: used here as a plant that is growing *in vitro,* or, as a verb, growing a plant *in vitro.*
cysteine (L-cysteine): an amino acid, a nitrogen-containing, organic compound.
cytogenetics: the study of chromosomes as applied to cellular activity and variability.
cytokinins: a group of growth regulators that induce bud formation and cell multiplication.

dedifferentiate: to revert to an undifferentiated state.
deionize: to remove ions from water by the use of a deionizer.
deionizer: a system that removes ions from water.
dichotomous growth: branching into two parts.
differentiate: to develop tissues or organs with specific functions.
diploid (2N): having two sets of chromosomes which is typical of vegetative (somatic) cells.

EDTA: a compound that makes iron in culture media more readily available to plants, a chelating agent, ethylenediamine tetraacetic acid.
element: a substance that cannot be separated into different substances by any usual chemical means.
embryo culture: culture of embryos excised from immature or mature seeds.
embryogenesis: formation of the embryo either in the seed (sexual) or in culture (somatic); here refers to production of embryoids in callus or cell culture.
embryoids: embryo-like vegetative structures developing in some cell and callus cultures, capable of developing into embryos which produce whole plants.
endosperm: the nutrient tissue or liquid surrounding a developing embryo in a seed.
enzyme: an organic catalyst.
epinasty: a downward bending of a leaf due to the more rapid growth of the upper surface.
etiolated: a state of growth induced by growing in absence of light and characterized as being white and elongated.
excise: to remove by cutting.
explant: the part of a plant used to start a culture.
exudate: matter discharged or diffused from a culture.

flaming: the sterile technique procedure of dipping an instrument in alcohol followed by setting it aflame to remove the alcohol and further sterilize the instrument.
f. c.: foot candle, a measure of illumination on one square foot of surface that is one foot away from a standard candle.

formula: (1) a chemical compound expressed in letters, numbers, or symbols to indicate its composition; (2) the ingredients, or recipe, for a tissue culture medium.

genetic: —variability, any variation in characteristics due to genetic inheritance or mutation. —engineering, the manipulation of genes and chromosomes to vary plant characteristics. —instability, the tendency of cells to mutate.

gibberellins: a group of growth regulators influencing cell enlargement.

growth regulators: organic compounds that influence growth and multiplication, such as cytokinins and auxins.

haploid: having half the normal number of chromosomes in vegetative cells. (N, or IN)

hardening off: the process of gradually accustoming plants to less humid conditions than experienced in culture or during rooting.

HEPA: "high efficiency particulate air" filters are an essential component of laminar air flow hoods.

herbaceous plants: seed plants without woody tissues.

heterotrophic: requiring organic substances for nutrients.

hood: see transfer chamber.

hormones: natural or synthetic chemicals that strongly affect growth, i.e., cytokinins, auxins, and gibberellins.

hot plate/stirrer: an electrical device that serves both as a hot plate and/or a magnetic stirrer. See magnetic stirrer.

hybrid: the plant from a cross between two different cultivars usually by sexual reproduction. See somatic hybrid.

hydrate: a compound with chemically bound water (water of hydration).

hypocotyl: that portion of a seedling stem below the cotyledons (first "leaves") and above the roots.

indexing: testing of plants for pathogens or contaminants.

indole-3-butyric acid: an auxin, abbreviated IBA.

inorganic chemicals: chemicals not designated as organic, usually without carbon.

inositol ($C_6H_6(OH)_6$): a member of the Vitamin B complex.

intergeneric: said of a cross between two different genera.

internode: the section of stem between two nodes.

interspecific: said of a cross between two separate species.

in vitro: literally, in glass (Latin). Used interchangeably with the terms tissue culture and micropropagation.

in vivo: occurring naturally, not in vitro.

ions: atoms or groups of atoms that carry a positive or negative charge.

juvenility: growth phase in a plant life cycle associated with young or immature state prior to its ability to flower.

laminar air flow: (see transfer chamber) even air flow directed in a specific, layered direction, without turbulence.

layer: a shoot or twig bent down and partly covered with soil so that it may take root.

magnetic stirrer: a device often combined with a hot plate in which a whirling magnet attracts a magnetized stir bar placed inside a glass container on the plate.

media: plural of medium.

medium: (1) liquid or agar solidified nutrient substance in or on which tissue cultures are grown; (2) soil mix.

meiosis: reduction division of chromosomes that occurs in the formation of pollen or the ovule nucleus resulting in half the normal number of chromosomes contained in vegetative cells.

meniscus: the curved upper surface of a column of liquid.

meristem: undifferentiated cells at the tips of roots and shoots, also found as cambium, and in other locations. The term is used loosely in this book to denote a microscopic shoot tip explant usually under 1.5 mm. The term is also used as a verb meaning to excise these small shoot tips or "meristems." A true meristem or meristematic dome is usually about .5 mm or less.

meristemoid: resembling meristem.

micromho: a millionth of a mho, a unit of electrical conductivity. (Mho is ohm spelled backwards and each is the reciprocal of the other).

micropropagation: extremely small propagation, used interchangeably with tissue culture or in vitro culture, but actually, more specific for describing multiplication in vitro.

mitosis: division and separation of chromosomes during vegetative cell division.

mole: molecular weight of a compound expressed in grams.

molecular weight: the sum of atomic weights in a molecule.

molecule: the smallest division of a compound that still retains the characteristics of that compound.

morphogenesis: beginning of form.

mutagenesis: the formation of mutants.

mutant: a deviation from the normal genetic arrangement.

mutation: the occurrence of a heritable variation in an individual due to a change in genes or chromosomes.

node: that part of a stem from which a leaf or shoot originates.

nucleic acids: any of a group of complex organic acids found especially in the nucleus of all living cells.

ohm: a unit of electrical resistivity, the reciprocal of a mho.

organic chemicals: refers to most compounds containing carbon.

organogenesis: beginning of organs, i.e. shoot or root.

osmosis: the passage of water from a less dense to a more dense solution through a membrane.

osmotic pressure: the pressure potential existing in the diffusion of a solvent through a semi-permeable membrane into a more concentrated solution in order to equalize the concentrations on both sides of the membrane.

pantothenic acid ($C_9H_{17}O_5N$): a member of the vitamin B complex.

parafilm: a stretchy, breathable, easily cut tape used to seal culture containers. It is readily available from scientific equipment supply companies.

parenchyma: tissue of relatively undifferentiated cells found in leaves and roots.

pathogen: disease causing microorganism.

petiole: a leaf stalk or stem.

pH: a symbol of the degree of acidity or alkalinity on a scale of 1–14.

phenolic: organic compounds often the by-products of metabolism and sometimes toxic. Phenolic exudate: that which is exuded or oozed out.

photoperiod: the length of time plants are exposed to light.

pipet: a slender glass tube used to suction small amounts of liquid.

pipetter: a device for pipetting.

plagiotropic: horizontal as opposed to vertical growth, usually with reference to conifers where the main stem does not grow upright.

plasmolysis: the separation of cytoplasm from the cell wall due to removal of water from the protoplast.
polysaccharide: a group of carbohydrates composed of many units of various sugars.
precipitate: a substance that separates out of solution.
premix: used here to depict media formulas prepared commercially in (usually) a dry form.
primordia: plural of primordium; plant organs in earliest stages of differentiation.
proliferation: rapid multiplication.
propagule: a small bit of plant that is being propagated, a transfer.
proteins: a large group of complex organic substances containing amino acids and diverse elements.
protoplast: a cell without a cell wall but with a membrane.
protoplast fusion: the uniting of two protoplasts.
pyridoxine ($C_8H_{11}O_3N$): Vitamin B_6.

reduction: a chemical process in which an electron is added to an atom or an ion.
reduction division: see meiosis.
regeneration: the production of new plants or parts of plants.
resin bed: with respect to water purification: a tank or cartridge containing charged resin particles to attract ions of opposite charge from mineral containing water.
resistivity: the electrical resistance between opposite faces of a unit cube of a substance; usually for a one-centimeter cube.
riboflavin: Vitamin B_2.
rotator: a rotating device to hold and agitate liquid culture containers.

scape: leafless flower stalk growing from the root crown.
secondary products: products of plant metabolism that are not primarily related to growth and reproduction, such as medicinals, flavorings, dyes, pesticides, etc.
senescent: aging.
significant figures, concept of: the concept of rounding off a series of numbers to some meaningful number of figures.
somatic: vegetative, as opposed to sexual. —hybrid: vegetatively derived hybrid.
somatic hybridization: the creation of hybrids by vegetative means, i.e., protoplast fusion.
sport: a mutation.
stages of culture: Stage I, establishment; Stage II, multiplication; Stage III, rooting.
sterile technique: the art of working with cultures in an environment free from microorganisms.
stock plants: plants from which other plants are started.
stock solutions: concentrated solutions from which portions are used to make media.
stomata: small openings in the epidermis of leaves and stems through which gases and water may pass. Normally these open and shut depending upon temperature and humidity.
symbiosis: two dissimilar organisms living together usually to mutual advantage.
synthesis: putting together parts or elements to make a whole.

tare: to allow for weight of a container.
tetraploid: double the normal number of chromosomes in vegetative cells.
thiamine ($C_{12}H_{17}ON_4SCl$): Vitamin B_1.
tissue culture: literally the culture of tissues, therefore, a misnomer for its broader use as a synonym for in vitro culture.

totipotence: the capability of developing into a whole plant, said of a cell.

transfer: the process of dividing cultures and placing the sections in containers of fresh sterile medium, or the pieces being transferred.

transfer chamber (hood): a largely enclosed area in which culture transfers are made using sterile technique.

ultrasonic cleaner: a cleaning device employing ultra sonic vibrations, sometimes used to clean explants.

undifferentiated: refers to cells or tissues that are not yet modified for their ultimate role, not differentiated.

vegetative: somatic, nonsexual.

vitrification: an undesirable phenomenon that sometimes develops in cultures, appearing as overly succulent, crisp, water-logged, or glassy tissues.

Bibliography

1. Abdullah, Anwar A., Michael M. Yeoman and John Grace, 1985. In vitro adventitious shoot formation from embryonic and cotyledonary tissues of *Pinus brutia* Ten. *Plant Cell Tissue Organ Culture* 5:35–44.
2. Almehdi, Ali A., and Dan E. Parfitt, 1986. In vitro propagation of peach: 1. Propagation of 'Lovell' and 'Nemagauard' peach rootstocks. *Fruit Varieties Jn.* 40(1):12–17.
3. Ammirato, P. V., 1983. Embryogenesis, pp. 82–123. In: D. A. Evans, W. R. Sharp, P. V. Ammirato, Y. Yamada (Eds). *Handbook of plant cell culture,* Vol. 1: Techniques for propagation and breeding. Macmillan, N.Y.
4. Ammirato, P. V., 1983. The regulation of somatic embryo development in plant cell culture, Vol. 1: Techniques for propagation and breeding. Macmillan, N.Y.
5. Anderson, W. C. and Kathryn Mielke, 1985. *Outline for Micropropagating Virus Free Bulbous Iris.* W.S.U., Northwestern Washington Research and Extension Center, 1468 Memorial Highway, Mt, Vernon, WA 98273.
6. Anderson, W. C., 1978. Rooting of tissue cultured rhododendrons. *Proc. Int. Plant Prop. Soc.* 28:135–139.
7. Anderson, W. C., 1984. Primula micropropagation multiplies special plants. *Primroses* Vol. 42, 2:21–23.
8. Anderson, W. C., and G. W. Meagher, 1978. *Cost of propagating plants through tissue culture using lilies as an example.* Oregon State University Ornamentals Short Course.
9. Anderson, W. C. and G. W. Meagher, 1977. Cost of propagating broccoli plants through tissue culture. *HortSci.* 12(6):543–544.
10. Anderson, W. C. and James B. Carstens, 1977. Tissue culture propagation of broccoli, *Brassica oleracea* (Italica Group), for use in F1 hybrid seed production *J. Am. Soc. Hort. Sci.* 102:(1):69–73.
11. Anderson, W. C. and Greg Haner, 1978. *Strawberry tissue culture formula.* Wash. State University, N.W. Wash. Res. Unit, Mount Vernon, Wash.
12. Anderson, W. C., 1977. Rapid propagation of *Lilium,* c.v. 'Red Carpet'. *In Vitro* 13(3):145 (Abstract).
13. Anderson, W. C., 1978. *Progress in tissue culture propagation of rhododendrons.* Ornamentals Short Course, Oregon State University, Portland, OR.
14. Anderson, W. C., 1975. Propagation of rhododendrons by tissue culture: Part 1. Development of a culture medium for multiplcation of shoots. *Proc. Inter. Plant Prop. Soc.* 25:129–135.
15. Arditti, J., 1977. Clonal propagation of orchids by means of tissue culture—a manual, In: J. Arditti (ed.) Orchid biology, reviews and perspectives, I. Cornell Univ. Press, Ithaca, N.Y.
16. Ball, E. A., 1978. Cloning in vitro of *Sequoia sempervirens.* In *Fourth Int'l. Cong. Plant Tissue and Cell Culture.* T. A. Thorpe, ed. Calgary, Canada. Abstract N. 1726:163
17. Barker, Pamela K., R. A. deFossard, and R. A. Bourne, 1977. Progress toward clonal propagation of eucalyptus species by tissue culture techniques. *Proc. Int. Plant Prop. Soc.* 27:546–556.

18. Barlass, M. and Skene, 1978. In vitro propagation of grapevine (*Vitis vinifera* L.) from fragmented shoot apices. *Vitis* 17:335–340.

19. Beaty, R. M., E. O. Franco, and O. J. Schwarz, 1985. Hormonally induced shoot development on cotyledonary explants of *Pinus occarpa. In Vitro,* V.21, 3, Part II:53A.

20. Bilkey, P. C., B. H. McCown, and A. C. Hildebrandt, 1978. Micropropagation of African violet petiole cross-sections. *HortSci.* 13(1):37–38.

21. Biondi, Stefania, and Trevor A. Thorpe, 1981. Clonal propagation of forest tree species. In *Proc. COSTED Symp. on Tissue Culture of Economically Important Plants,* An N. Rao, ed., Singapore. 197–204.

22. Borgman, C. A. and K. W. Mudge, 1986. Factors affecting the establishment and maintenance of 'Titan' red raspberry root organ cultures. *Plant Cell Tissue Organ Culture,* Vol. 6, 2:127.

23. Bottino, P. J., 1981. *Methods in Plant Tissue Culture.* Botany Dept., Univ. of Maryland, College Park, MD. Pub. by Kemtec Educational Corp.

24. Boulay, M. 1979. Multiplication et clonage rapide du *Sequoia sempervirens* par la culture "in vitro". AFOCEL Etudes et Recherches, *Micropropagation d'Arbres Forestiers,* 12:49–54.

25. Boxus, P., M. Quoirin, and J. M. Laine, 1977. Large scale propagation of strawberry plants from tissue culture. *Applied and Fundamental Aspects of Plant Cell, Tissue and Organ Culture.* J. Reinert and Y. P. S. Bajaj, eds Springer-Verlan, N.Y. 130–143.

26. Bridgen, Mark, 1986. Do-it-yourself cloning. *Greenhouse Grower,* May, 1986:43–47.

27. Briggs, Bruce, Briggs Nursery, Olympia, WA. Personal communication.

28. Chee, Raymond, Robert M. Pool, and Donald Bucher, 1984. A method for large scale in vitro propagation of *Vitis. New York's Food and Life Sciences Bulletin,* No. 109. N.Y. State Ag. Expt. Sta., Geneva, N.Y.

29. Chin, Chee-Kok, 1982. Promotion of shoot and root formation in asparagus in vitro by ancymidol. *HortSci.* 17(4):590–591.

30. Collins, G. B., and A. D. Genovesi, 1982. Anther culture and its application to crop improvements. In: *Application of Plant Cell and Tissue Culture to Agriculture and Industry.* D. T. Tomes et al, eds. Plant Cell Culture Centre, Univ. of Guelph, Guelph, Ontario, Canada.

31. Cooke, Ron C., 1977. Tissue culture propagation of African violets. *HortSci.* 12(6):549.

32. Cooke, Ron C., 1977. The use of an agar substitute in the initial growth of Boston ferns in vitro. *HortSci.* 12(4):339.

33. Cooke, Ron C. 1979. Homogenization as an aid in tissue propagation of *Platycerium* and *Davallia. HortSci.* 14:21.

34. Davies, D. R., 1980. Rapid propagation of roses in vitro. *Scientia Horticulturae* 13:385–389.

35. Davis, M. J., Ralph Baker, and Joe J. Hanan, 1977. Clonal multiplication of carnations by micropropagation. *J. Amer. Soc. Hort. Sci.* 102(1):48–53.

36. deFossard, R. A., 1976. Tissue culture propagation of *Eucalyptus ficifolia* F. Muell. *Proc. Int. Plant Prop. Soc.* 26:373–378.

37. deFossard, R. A., 1976. *Tissue Culture for Plant Propagators,* Univ. of New England, Armidale, New South Wales.

38. Dicosmo, F., 1985. Microbial insult and plant cell cultures: interactions with novel biotechnological potential. *In vitro* 21(3):60A. Abstract.

39. Dixon, R. A., ed., 1985. *Plant Cell Culture, a Practical Approach.* IRL Press, Oxford—Washington, D.C.
40. Donnelly, Danielle J., William E. Vidaver and Kwai Y. Lee, 1985. The anatomy of tissue cultured red raspberry prior to and after transfer to soil. *Plant Cell Tissue Organ Culture* 4:43–50.
41. Driver, J. A., 1985. Direct field rooting and acclimatization of tissue culture cuttings. T.C.A. annual meeting abstracts, *In Vitro* V.21, 3:57A, 80.
42. Dunstan, David I., 1981. Transplantation and post-transplantation of micropropagated tree-fruit rootstocks. *Proc. Int. Plant Prop. Soc.* 31:39–44.
43. Earle, E. D. and R. W. Langhans, 1975. Carnation propagation from shoot tip culture in liquid medium. *HortSci.* 10:608–610.
44. Feliciano, Ascunia J., and M. de Assis, 1983. In vitro rooting of shoots from embryo-cultured peach seedlings. *HortSci.* 18(5):705–706.
45. Fitter, Mindy, and A. D. Krikorian, 1985. Mature phenotype in *Hemerocallis* plantlets fortuitously generated in vitro. *J. Plant Physiol.* Vol. 121:97–101.
46. Fujita, Y., 1985. Efficient production of shikonin derivatives by cell suspension culture of *Lithospermum erythrorhizon. In Vitro* 21(3):60A. Abstract.
47. Gamborg, O. L., T. Murashige, T. A. Thorpe, and I. K. Vasil, 1976. Plant tissue culture media. *In Vitro* 12:473–478.
48. Gamborg, O. L., 1986. Cells, Protoplasts, and Plant Regeneration in Culture. In: Manual of Industrial Microbiology and Biotechnology. A. L Demain and N. A. Solomon, editors, Dept of Nutrition & Food Science, Mass. Inst. of Tech., Cambridge, Mass.
49. Gautheret, R. J., 1942. *Manuel technique de culture des tisus végétaux.* Massonet Cie, Paris.
50. Gautheret, R. J., 1982. Plant tissue culture: the history. In *Plant Tissue Culture 1982*—Proceedings of the 5th International Congress of Plant Tissue and Cell Culture held at Tokyo and Lake Yamanaka, Japan, July 11–16, 1982. Pages 1–10. Akio Fujiwara, ed. Pub. by the Japanese Association for Plant Tissue Culture. Dist. by Maruzen Co., Ltd., P.O. Box 5050, Tokyo International, 100–31, Japan.
51. Gebhardt K., 1985. Development of a sterile cultivation system for rooting of shoot tip cultures (red raspberries) in duraplast foam. *Plant Sci.* 39:141–148.
52. Goodwin, P. B. and T. Adisarwanto, 1980. Propagation of potato by shoot tip culture in petri dishes. *Potato Res.* 23:445–448.
53. Gould, Jean H., and Toshio Murashige, 1985. Morphogenic substances released by plant tissue cultures; 1. Identification of berberine in *Nandina* culture medium, morphogenesis, and factors influencing accumulation. *Plant Cell Tissue Organ Culture* 4:29–42.
54. Goyal, Yashpal, R. L. Bingham, and Peter Felker, 1985. Propagation of the tropical tree, Leucaena leucocephala K67, by in vitro bud culture. *Plant Cell Tissue Organ Culture,* Vol. 4, 1:3–10.
55. Griffis, John L. Jr., Gary Hennen, and Raymond P. Oglesby, 1980. Establishing tissue-cultured plants in soil. *Proc. Int. Plant Prop. Soc.* 33:618–622.
56. Gunter, Dan L., 1979. Plant cost estimation: the south Florida foliage case. *Proc. Int. Plant Prop. Soc.* 29:525–532.
57. Hammerschlag, F., 1982. Factors affecting establishment and growth of peach shoots in vitro. *HortSci.* 17(1):85–86.

58. Handley, Lewis W. and O. L. Chambliss, 1979. In vitro propagation of *Cucumis sativus*. *HortSci.* 14:22.

59. Harris, R. E., 1984. Rapid propagation of saskatoon plants in vitro. Ag. Res. Coun. of Alberta, Farming for the Future Project #83-0054.

60. Harris, R. E., 1985. Rooting of in vitro shoots of saskatoon (Amelancher alnifolia). Ag. Res. Coun. of Alberta, Farming for the Future Project #82-0017.

61. Harris R. E., and J. H. Stevenson, 1982. In vitro propagation of *Vitis*. *Vitis* 21:22–32.

62. Harris, R. E., and E. B. B. Mason, 1983. Two machines for in vitro propagation of plants in liquid media. *Can. Jn. Plant Sci.* 63:311–316.

63. Hartmann, Hudson T., William J. Flocker, and Anton M. Kofranek, 1981. *Plant Science*. Prentice-Hall, Inc., Englewood Cliffs, N.J.

64. Hartmann, Hudson T., and D. K. Kester, 1983. *Plant Propagation—Principles and Practices*. 4th Ed. Prentice-Hall, Inc., Englewood Cliffs, N.J.

65. Hartney, V. J., and E. D. Eabay, 1984. From tissue culture to forest trees. Proc. Intl. Plt. Prop. Soc. 34:93–99.

66. Hartney, V. J., 1982. Tissue Culture of Eucalyptus. Proc. Int. Prop. Soc. 32:98–109.

67. Hasagawa, Paul M., 1979. In vitro propagation of rose. *HortSci.* 14:610–612.

68. Hasagawa, Paul M., 1980. Factors affecting shoot and root initiation from cultured rose shoot tips. *J. Amer. Soc. Hort. Sci.* 105(2):216–220.

69. Hennen, G. R. and T. J. Sheehan, 1978. In vitro propagation of *Platycerium stemaria* (Beauvois) Deav. *HortSci.* 13(3):245.

70. Heuser, Charles W., and John Harker, 1976. Tissue culture propagation of daylilies. *Proc. Int. Plant Prop. Soc.* 26:269–272.

71. Heuser, C. W. and D. A. Apps, 1976. In vitro plantlet formation from flower petal explants of *Hemerocallis* cv. 'Chipper Cherry'. *Can. J. Bot.* 54:616–618.

72. Hildebrandt, A. C., A. J. Riker, and B. M. Duggar, 1946. The influence of the composition of the medium on growth in vitro of excised tobacco and sunflower tissue cultures. *Amer. J. Bot.* 33:591–597.

73. Hussey, G., 1976. Propagation of Dutch iris by tissue culture. *Scientia Horticulturae* 4:163–165.

74. Hussey, G., 1975. Totipotency in tissue culture explants and callus of some members of the Liliaceae, Iridaceae, and Amaryllidaceae. *Jn. Exp. Bot.* 216:253–262.

75. Johnson, J. L. and E. R. Emino, 1979. In vitro propagation of *Mammillaria elongata*. *HortSci.* 14(5):605–606.

76. Kao, K. N. and M. R. Michayluk, 1974. A method for high frequency intergeneric fusion of plant protoplasts. *Planta* 115:355–367.

77. Kasperbauer, M. J., R. C. Buckner, and L. P. Bush, 1979. Tissue culture of annual ryegrass × tall fescue F1 hybrids; callus establishment and plant regeneration. *Crop Sci.* 19:457–460.

78. Knudson, L., 1946. A new nutrient solution for the germination of orchid seeds. *Amer. Orchid Soc. Bull.* 15:214–217.

79. Krul, W. R. and J. Myerson, 1980. In vitro propagation of grape. In *Proc. of the Conf. on Nursery Prod. of Fruit Plants through Tissue Culture—applications and feasibility.* USDA-SEA Agri. Res. Results. Belltsvile.

80. Kunisaki, J. T., 1980. In vitro propagation of *Anthurium andreanum* Lind. *HortSci.* 15(4):508–509.

81. Kyte, Lydiane and Robert M. Kyte, 1981. Small fruit culture after the test tube. *Proc. Int. Plant Prop. Soc.* 31:45–47.

82. Kyte, Lydiane and Bruce Briggs, 1979. A simplified entry into tissue culture production of rhododendrons. *Proc. Int. Plant Prop. Soc.* 29:90–95.

83. Lane, David W., 1982. *Apple Trees in Test Tubes.* Agriculture Canada Research Station, Summerland, B.C.

84. Lazarte, J. E., M. S. Galser, and O. R. Brown, 1982. In vitro propagation of *Epiphyllum chrysocardium. HortSci.* 17(1):84.

85. Levy, Luis W., 1981. A large-scale application of tissue culture: The mass propagation of pyrethrum clones in Ecuador. *Env. and Exptl. Bot.,* Vol. 21, No. 3/4:389–395.

86. Linsmaier, Elfriede M., and Folke Skoog, 1965. Organic growth factor requirements of tobacco tissue cultures. *Physiol. Plant.* 18:100–128.

87. Lloyd, Gregory and Brent McCown, 1980. Commercially-feasisble micropropagation of mountain laurel, *Kalmia latifolia,* by use of shoot-tip culture. *Proc. Int. Plant Prop. Soc.* 30:421–427.

88. Loescher, Wayne H. and Carolyin Albrecht, 1979. Deveopment in vitro of *Nephrolepsis exaltata* cv. 'Bostoniensis' runner tissues. *Physiol. Plant.* 47:250–254.

89. Makino, R. K. and Makino, P. J., 1977. Propagation of *Syngonium podophyllum cultivars* through tissue culture. *In Vitro* 13:357.

90. Matsuyama, J. 1978. *Tissue culture propagation of nandina.* T.C.A. 29th Annual Meeting, Denver, p. 357. (Abstract).

91. Mauseth, J. 1979. A new method for the propagation of cacti. *Cactus & Succulents J.* 57:186–187.

92. McCown, Brent, and Ron Amos, 1979. Initial trials with commercial micropropagation of birch selections. *Proc. Int. Plant Prop. Soc.* 29:387–393.

93. McCullouch, Steven M. and Bruce A. Briggs, 1982. Preparation of plants for micropropagation. *Proc. Int. Plant Prop. Soc.* 32:297–303.

94. McRae, Edward A. and Judith F. McRae, 1979. Eight years of adventure in embryo culturing. In: *The Lily Yearbook of the North American Lily Society.*

95. Mele, E., J. Messeguer, and P. Camprubi, 1982. Effect of ethylene on carnation explants grown in sealed vessels. Proc. 5th Int'l Cong. Plant Tissue and Cell Culture. *Plant Tissue Culture,* 1982:69–70.

96. *Merck Index, an Encyclopedia of Chemicals and Drugs, 1976,* 9th edition. Martha Windholz, ed. Merck & Co., Inc. publisher, Rahway, N.J.

97. Meyer, M. M. Jr., 1976. Propagation of daylilies by tissue culture. *HortSci.* 11(5):485–487.

98. Meyer, M. M. Jr., L. H. Fuchigami and A. N. Roberts, 1975. Propagation of tall bearded iris by tissue culture. *HortSci.* Vol. 10(5):497–480.

99. Meyer, Martin M. Jr., 1980. In vitro propagation of *Hosta sieboldiana. HortSci.* Vol. 15(6):737–738.

100. Meyer, Martin M. Jr., 1982. In vitro propagation of *Rhododendron catawbiense* from flower buds. *HortSci.* Vol. 17(6):891–892.

101. Mielke, Kathryn Ann, 1984. *In Vitro Bulbing of Bulbous Iris CV 'Blue Ribbon'* ***(Iris xiphium × Iris tingitana)***. Master's degree thesis, Wash. State Univ., Pullman, WA.

102. Mii, M., T. Mori, and N. Iwase, 1974. Organ formation from the excised scales of *Hippeastrum hybridum* in vitro. *J. Hort. Sci.* 49:241–244.

103. Mikkelsen, Edward P. and Kenneth C. Sink, Jr., 1978. In vitro propagation of Rieger Elatior begonias. *HortSci.* 13:(3):242–244.

104. Miller, G. A., D. C. Coston, E. G. Denny, and M. E. Romeo, 1982. In vitro propagation of 'Nemaguard' peach rootstock. *HortSCi.* 17(2):194.

105. Monette, Paul L., 1983. Influence of size of culture vessel on in vitro proliferation of grape in a liquid medium. *Plant Cell Tissue Organ Culture* 2:327–332.

106. Monette, Paul L., 1986. Micropropagation of kiwifruit using non-axenic shoot tips. *Plant Cell Tissue Organ Culture* 6:73–82.

107. Morel, G. and J. F. Muller, 1964. *La culture in vitro du méristème apical de la pomme de terre.* C.R. Acad. Sc. Paris 258:5250–5252.

108. Morel, G. M., 1974. Clonal multiplication of orchids. In: *The Orchids: Scientific Studies.* C. L. Withner, ed., John Wiley and Sons, N.Y., 169–222.

109. Morel, George M., 1964. Tissue culture—A new means of clonal propagation of orchids. *Am. Orch. Soc. Bull.* 33:473–478.

110. Murashige, T., M. N. Shaba, Paul A. Hasagawa, F. H. Taktori, and J. B. Jones, 1972. Propagation of asparagus through shoot apex culture. 1. Nutrient medium for formation of plantlets. *Jn Amer. Soc. Hort. Sci.* 97:158–161.

111. Murashige, T., 1974. Plant Propagation through tissue cultures. *Ann. Rev. Plant. Physiol.* 25:135–166.

112. Murashige, Toshio, Maria Serpa, and Jeanne B. Jones, 1974. Clonal multiplication of *Gerbera* through tissue culture. *HortSci.* 9(3):175–180.

113. Murashige, T. and F. Skoog, 1962. A revised medium for rapid growth and bioassays with tobacco tissue cultures. *Physiol. Plant.* 15:473–497.

114. Nag, K. K., and H. E. Street, 1973. Carrot embryogenesis from frozen cultured cells. *Nature* 245:270–272.

115. Nickerson, N. L., 1978. In vitro shoot formation in lowbush blueberry seedling explants. *HortSci.* 13(6):698.

116. Nitsch, J. P. and C. Nitsch, 1969. Haploid plants from pollen grains. *Science* 163:85–87.

117. Oki, Lorence R., 1981. The modification of research procedures for commercial propagation of Boston ferns. *Env. and Exp. Bot.* Vol. 21, No. 3/4:397–400.

118. Page, Y. M., and J. Van Staden, 1984. In vitro propagation of Hypoxis rooperi. *Plant, Cell, Tissue Organ Culture,* Vol. 3, 4:359–362.

119. Papachatzi, Marietta, P. Allen Hammer and Paul M. Hasegawa, 1980. Tissue culture propagation of *Hosta decorata* "Thomas Hogg." *HortSci.* 15(3):436.

120. Papachatzi, Marietta, P. Allen Hammer, and Paul M. Hasegawa, 1980. In vitro propagation of *Hosta plantaginea. HortSci* 15(4):506–507.

121. Parfitt. Dan E., and Ali A. Almehdi, 1986. In vitro propagation of peach: II. A medium for in vitro multiplication of 56 cultivars. *Fruit Varieties Jn.* 40(2):46–47.

122. Peterson, Marcus A., 1979. Some aspects of nursery production in Queensland. *Proc. Int. Plant Soc.* 29:109–110.

123. Pierik, R. L. M., H. H. M. Steegmans, and J. A. J. Vak de Meys, 1974. Plantlet formation in callus tissues of *Anthurium andreanum* Lind. *Sci. Horticul.* 2:193–198.

124. Pierik, R. L. M. and Th. A. Segers, 1973. In vitro culture of midrib explants of *Gerbera:* Adventitious root formation and callus induction. *Z. Pjlanzenphysiol Bd.* 69.S.204–212.

125. Read, Paul E., Stephen Garton, K. A. Louis, E. S. Zimmerman, 1982. In vitro propagation of species for bioenergy plantations. In *Plant Tissue Culture 1982,* Proc. 5th Int'l Cong. Plant Tissue and Cell Culture, Akio Fujiweara, ed. pp. 757–758.

126. Read, Paul E. and P. Gavinlertvatana, 1976. Advances in tissue culture: Ray flower and protoplast culture. *Proc. Int. Plant Prop. Soc.*

127. Rechcigl, Miloslav, Jr., Ed. in chief, *CRC Handbook Series in Nutrition and Food,* Section G, Vol. III, 1978. CRC Press, Inc., Cleveland, Ohio.

128. Roest, S. and C. S. Bokelmann, 1975. Vegetative propagation of *Chrysanthemum cinerariaefolium* in vitro. *Sci. Hort.* 1:120–122.

129. Sagawa, Koneo, and John T. Kunisaki, 1982. Clonal propagation of orchids by tissue culture. *Plant Tissue Culture 1982,* Proc. 5th Int'l Cong. Plant Tisue and Cell Culture, Akio Fujiwara, ed. pp. 683–684.

130. Salisbury, Frank B. and Cleon Ross, 1969. *Plant Physiology.* Wadsworth Pub. Co., Inc., Belmont, CA.

131. Schenk, R. U., and A. C. Hildebrandt, 1972. Medium and techniques for induction and growth of monocotyledonous and dicotyledonous plant cell cultures. *Can. Jn. Bot.* 50:199–204.

132. Schnabelrauch, L. S. and K. C. Sink, 1979. In vitro propagation of *Phlox subulata* and *Phlox paniculata. HortSci.* 14(5):607–608.

133. Scoog, F. and C. O. Miller, 1957. Chemical regulation of growth and organ formation in plant tissues cultured in vitro. *Symp. Soc. Exp. Biol.,* 11:118–130.

134. Seabrook, J. E. A. and B. G. Cumming, 1977. The in vitro propagation of amaryllis (*Hippiastrum* spp.) hybrids. *In Vitro,* Vol. 13:831–836.

135. Skolmen, Roger G. and Marion O. Mapes, 1978. Aftercare procedures required for field survival of tissue culture propagated *Acacia koa. Proc. Int. Plant Prop. Soc.* 28:156–164.

136. Smith, R. H. and A. E. Nightingale, 1979. In vitro propagation of *Kalanchoe. HortSci.* 14(1):20.

137. Smith, William A., 1981. The aftermath of the test tube in tissue culture. Proc. *Int. Plant Prop. Soc.* 31:47–49.

138. Sommer, Harry E., and Linda S. Caldas, 1981. In vitro methods applied to forest trees. In *Plant Tissue Culture, Methods and Applications in Agriculture* Trevor A. Thorpe, ed., Academic Press.

139. Song, John S., 1982. *An inside look at some economics concerning plant tissue culture lab facilities and operation.* Magenta Corp., Chicago, Ill.

140. Stamets, Paul, and J. S. Chilton, 1983. *The Mushroom Cultivator, A Practical Guide to Growing Mushrooms at Home.* Agarikon Press, Olympia, Washington.

141. Steward, F. C. and A. D. Krikorian, 1971. *Plants, Chemicals and Growth.* Academic Press.

142. Stimart, Dennis P., and Peter D. Ascher, 1981. Foliar emergence from bulblets of *Lilium longiflorum* Thumb. as related to in vitro generation temperatures. *J. Amer. Soc. Hort. Sci.* 106(4):446–450.

143. Stimart, Dennis P., and Peter D. Ascher, 1981. Developmental responses of *Lilium longiflorum* bulblets to constant or alternating temperatures in vitro. *J. Amer. Soc. Hort. Sci.* 106(4):450–454.

144. Stolz, Leonand P., 1979. Getting started in tissue culture—equipment and costs. *Proc. Int. Plant Prop. Soc.* 29:375–381.

145. Suttle, Gayle R. L., 1983. Micropropagation of deciduous trees. *Proc. Int. Plant Prop. Soc.* 33:46–49.

146. Suttle, Gayle, Microplant Nurseries, Inc., Gervais, OR. Personal communication.

147. Takayama, S. and M. Misawa, 1980. Differentiation in *Lilium* bulbscales grown in vitro. Effects of activated charcoal, physiological age of bulbs and sucrose concentration on differentiation and scale leaf formation in vitro. *Physiol. Plant.* 48:121–125.

148. Takayama, Shinsaku, Masanaru Misawa, Yoshiki Takashige, and Hiroshi Tsumori, 1982. Cultivation of in vitro propagated *Lilium* bulbs in soil. *J. Amer. Soc. Hort. Sci.* 107(5):830–834.

149. Thomas, Donovan des S., and Toshio Murashige, 1979. Volatile emissions of plant tissue cultures, I. Identification of the major components. *In Vitro* 15(9):654–662.

150. Thompson. D. C. and J. B. Zaerr. 1981. *Induction of adventitious buds on cultured shoot tips of Douglas-fir. (**Pseudotsuga menziesii** (Mirb) Franco).* From IUFRO Colloque International sur la Culture "in vitro" des essences forestières, AFOCEL, Nangis, France.

151. Thrope, Trevor A., Ed., 1981. *Plant Tissue Culture—Methods and Applications in Agriculture.* Academic Press.

152. Tisserat, B., 1985. Embryogenesis, Organogenesis and Plant Regeneration. In *Plant Cell Culture: a Practical Approach,* R. A. Dixon, ed. IRL Press Ltd., P.O. Box 1, Eynsham, Oxford OXY 1JJ, England.

153. Tomes, D. T., B. E. Ellis, P. M. Harney, K. J. Kasha, and R. L. Peterson, editors, 1982. *Application of Plant Cell and Tissue Culture to Agriculture and Industry.* Plant Cell Culture Centre, University of Guelph, Guelph, Ontario, Canada.

154. Torello, William AS. and A. G. Symington, 1984. Regeneration from perennial rye grass callus tissue. *HortSci.* 10(1):56–57.

155. Upham, Sarah, 1983. Native Plants, Plant Resources Inst., 360 Wakara Way, Salt Lake City, Utah 84108. Personal communication.

156. Vacin, E. F and F. W. Went, 1949. Some pH changes in nutrient solutions. *Bot. Gaz.* 110:605–613.

157. Van Aartrijk and P. C. G. van der Linde, 1986. In vitro propagation of flower-bulb crops. In: *Tissue Culture as a Plant Production System for Horticultural Crops.* R. H. Zimmerman, R. J. Griesbach, F. A. Hammerschlag, R. H. Lawson, editors, USDA. Martinus Nijhoff Pub.

158. Vasil, I. K. (ed.), 1984. Cell Culture and Somatic Cell Genetics in Plants. Laboratory techniques. Academic Press, Inc., New York.

159. Von Arnold, Sara, and Tage Eriksson, 1981. In vitro studies of adventitious shoot formation in *Pinus contorta. Can. Jn. Bot.* V59:870–872.

160. Waldren, Doug, 1985. Data Acquisition Manual, Remote Ecological Environments. Weyerhaeuser Co.

161. Wehner, Todd. and Robert D. Locy, 1981. In vitro adventitious shoot and root formation of cultivars and lines of *Cucumis sativus* L. *HortSci* 16(6):759–760.

162. Wetherell, D. F., 1982. *Introduction to in vitro propagation.*Avery Pub. Group Inc., Wayne, N.J.

163. White, P. R., 1963. *The Cultivation of Animal and Plant Cells.* 2nd Ed., Ronald Press Co., N.Y.

164. White, P. R., 1943. *A Handbook of Plant Tissue Culture.* Jacque Cattell Press, Inc., Tempe, Ariz.

165. Wilkins, Malcolm B., 1984. *Advanced Plant Physiology.* Pitman Publishing Limited, London, England.

166. Williams, Margot, *Tissue culture of diploid **Hippeastrum reginae** Herb.* USDA, Sci. and Ed. Admin., Ag. Res., U.S. Nat'l Arboretum Wash, D.C. 20002.

167. Wimber, Donald E., 1963. Clonal multiplication of cymbidiums through tissue culture of the shoot meristem. *Am. Orch. Soc. Bull.* 32:105–107.

168. Withers, Lyndsey A. and J. T. Williams, 1986. *IBPGR Research Highlights—In Vitro Conservation.* IBPGR, FAO, Rome, Italy.

169. Wong, Steve, 1981. Direct rooting of tissue cultured rhododendrons into an artificial soil mix. *Proc. Int. Plant Prop. Soc.* 31:36–37.

170. Yang, H. J. and W. J. Clore, 1973. Rapid vegetative propagation of asparagus through lateral bud culture. *HortSci.* 8:141–143.

171. Yang, H. J. and W. J. Clore, 1974. Development of complete plantlets from vigorous shoots of stock plants of asparagus in vitro. *HortSci.* 9:138–140.

172. Young, Peter M., Anita Hutchins, and Marilyn L. Canfield, 1984. Use of antibiotics to control bacteria in shoot cultures of woody plants. *Plant Sci. Ltrs.* 34:203–209.

173. Zilis, Mark, Douglas Zwagerman, David Lamberts, and Lawrence Kurtz, 1979. Commercial propagation of herbaceous perennials by tissue culture. *Proc. Int. Plant Prop. Soc.* 29:404–413.

174. Zillis, Mark, and Douglas Zwagerman, 1979. Clonal propagation of hostas by scape section culture in vitro. *HortSci.* 14(3):80.

175. Zimmerman, Richard H., 1978. Tissue culture of fruit trees and other fruit plants. *Proc. Int. Plant Prop. Soc.* 28:539–546.

176. Zimmerman, Richard H. and Olivia C. Broome, 1980. Micropropagation of thornless blackberry. *Proc. Conf. on Nursery Prod. of Fruit Plants through Tissue Culture—Applications and Feasibility,* pp. 23–26. USDA-SEA Ag. Res. Results. ARR-NE-11.

177. Zimmerman, Richard H. and Olivia C. Broome, 1980. Blueberry micropropagation. *Proc. Conf. on Nursery Prod. of Fruit Plants through Tissue Culture—Applications and Feasibility* pp. 44–47. USDA-SEA Ag. Res. Results. ARR-NE-11.

178. Zimmerman, Richard H. and Olivia C. Broome, 1980. Apple cultivar micropropagation. *Proc. of the Conf. on Nursery Prod. of Fruit Plants through Tissue Culture—Applications and Feasibility,* pp. 54–58. USDA-SEA Ag. Res. Results. ARR-NE-11.

179. Zimmerman, R. H., R. J. Griesbach, F. A. Hammerschlag, and R. H. Lawson, editors, 1986. "Tissue Culture as a Plant Production System for Horticultural Crops." USDA. Martinus Nijhoff Pub.

Index